Majestic Universe, Views from Here to Infinity

Witness
the birth of stars; scan the
sky for extrasolar planets; venture
near black holes; travel into the realm of
galaxies and clusters of galaxies; glimpse the
ultimate limits of space thanks to the Hubble Deep
Field; gauge the vastness of space with Hipparcos; and
finally wonder about the history and future of our
mysterious universe. This beautiful large-format book takes us
on an exploration of the vast expanse of the universe through
the magic of more than 200 incredible photographs taken by
the Hubble Space Telescope and the largest telescopes in the
world. After describing the great theories of the structure of
the cosmos, Brunier describes the latest advances in
cosmology in terms that are accessible to everyone.
Probing deeper and deeper into space, he delves
farther and farther back in time seeking the
answer to the question of the origin of
the universe, of space, and
of time.

Serge Brunier

Majestic Universe
Views from Here to Infinity

Translated by Storm Dunlop

PUBLISHED BY THE PRESS SYNDICATE OF THE UNIVERSITY OF CAMBRIDGE
The Pitt Building, Trumpington Street, Cambridge, United Kingdom

CAMBRIDGE UNIVERSITY PRESS
The Edinburgh Building, Cambridge CB2 2RU, UK   www.cup.cam.ac.uk
40 West 20th Street, New York NY 10011-4211, USA   www.cup.org
10 Stamford Road, Oakleigh, Melbourne 3166, Australia
Ruiz de Alarcón 13, 28014 Madrid, Spain

First published in French as *l'Univers*, by Serge Brunier 1998
First published in English 1999

Printed in Italy by G. Canale C. & S.p.a. – Borgaro T.se – Turin, Italy

Typeface Zapf Humanist 9/11pt. *System* QuarkXpress® [HM]

A catalogue record for this book is available from the British Library

ISBN 0 521 66307 5 hardback

Contents

PREFACE

t was mirages that inspired this book – rather specific mirages, that are not created over the burning sands of the desert, but in the freezing depths of space. These mirages, which appear as arcs, circles, and crosses in the sky, are probably one of the most important scientific discoveries of recent decades. These images, produced by gravitational lenses, are the multiple, deformed, magnified, or amplified reflections of galaxies that existed more than ten billion years ago. They are one of the most astounding proofs of the validity of the models of the universe that astrophysicists have built up since the birth of scientific cosmology, some sixty years ago. This is because they prove that Albert Einstein was right: we do really live in a curved space-time.

The aim of this book is to present the reader with the account of a wonderful decade in which – quite apart from the discovery of the cosmic lenses – the spatial and temporal limits of the cosmos have been pushed back practically as far as the Big Bang, that strange, inaccessible instant in the history of the universe. Cosmology – the discipline that aims to study the universe as a whole – has been, for a long time, the poor relation when it comes to the atlases and encyclopaedias of astronomy. Cosmological theories were hardly mentioned, far less illustrated, for the lack of suitable material. Telescopes were simply unable to delve sufficiently far back in time. For some years, however, the most powerful astronomical instruments in the

world, the Hubble Space Telescope, of course, but also Hipparcos, ISO, Rosat, and Compton in space, and the giant telescopes on Hawaii, in the Andes, in New Mexico, and elsewhere, have literally overturned the idea we had of the world. They are now showing us events that occurred 5, 10, and even 12 billion years ago.

These images bear witness, above everything, to how the universe has changed over the course of time, in accordance with the predictions of the Big-Bang theory. The notion of the evolution of the universe as a whole represents a major break with previous philosophy. Scientists have been prepared for it for some decades, they know now that it is no longer just a hypothesis, but a fact.

In this book we have therefore given the distant cosmos more than its due, as well as the most recent investigations concerning the origin and structure of the universe. The reader will find here, often for the first time, the breathtaking images of the sky that show the reality of the curvature of space-time and of the evolution of the universe.

Authors are sometimes prejudiced. It is true that in this book, the reader will search in vain for details of the so-called alternative astrophysical theories to the Big Bang. Once again, this choice is explained by the profusion of observations obtained by astronomers in recent years, which all support the theories of an expanding universe, based on the equations of general relativity.

Books are often a matter of a meeting of minds. My thanks are

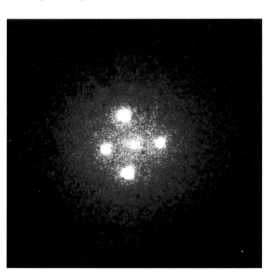

■ THE EINSTEIN CROSS. THIS QUADRUPLE IMAGE OF A DISTANT QUASAR SURROUNDING A CLOSER GALAXY IS AN ILLUSION – A MIRAGE – CAUSED BY THE CURVATURE OF SPACE-TIME. THE PHENOMENON WAS PREDICTED BY THE THEORY OF GENERAL RELATIVITY.

due to Dominique Wahiche and Séverine Cuzin, who have helped me to undertake this project, to improve it, and see it to completion. I also thank Gilles Seegmuller, the designer of this book's layout, who took a new look at material that was sometimes rather abstract, only to present it to such advantage. Thanks, finally to my partner Christine Maveyraud, who so patiently encouraged me throughout the three years that it took for this work to see fruition.

There would not be any books about astronomy without astronomers. I have worked alongside them for some twenty years. This book, of course, is inspired by their most recent work, published is highly technical scientific journals, such as the *Astrophysical Journal* or *Astronomy and Astrophysics*. Far more valuable, however, have been those discussions, over a cup of steaming coffee or tea, drunk in the control room of a giant telescope, during a sleepless night passed under a dome at the observatory at La Silla or on Mauna Kea, with researchers who were struggling with a recalcitrant electronic camera, a capricious sky, or an invisible quasar…

This book is, in fact, dedicated to astronomers, to their thirst for exploration, and to their passion for communication: thanks to Alain Blanchard, Jean-Marc Bonnet-Bidaud, Mark Dickinson, Olivier Le Fèvre, Marijn Franx, François Hammer, Anne-Marie Lagrange, Jean-Paul Knieb, David Malin, Yannick Mellier, Laurent Nottale, Francesco Paresce, and all those that I have happened to meet, at dusk, with a torch in their hands, on their way to their dome to pass the night searching for stars and galaxies.

Serge Brunier

A history of cosmology

■ BEFORE THE RENAISSANCE, THE UNIVERSE CONSISTED OF
NOTHING MORE THAN THE PLANETS OF THE SOLAR SYSTEM,
REVOLVING INSIDE THE STAR-STUDDED SPHERE OF THE HEAVENS.
LATER, ASTRONOMERS BELIEVED THAT THE MILKY WAY, AND ITS
MYRIADS OF STARS CONSTITUTED THE WHOLE COSMOS. WE OWE
OUR CURRENT CONCEPT OF THE UNIVERSE TO THE ASTRONOMERS
AND PHYSICISTS OF THE EARLY 20TH CENTURY. HERE, THE
MILKY WAY APPEARS ABOVE THE SAHARA DESERT, WITH THE
BEAUTIFUL CONSTELLATION OF THE SOUTHERN CROSS
APPEARING AS IF IT HAD BEEN STUCK INTO THE SAND.

■ WHEN COMETS CROSSED THE HEAVENS IN ANTIQUITY, THEY DISTURBED THE HARMONY OF THE SPHERES, BECAUSE PHILOSOPHERS WERE UNABLE TO AGREE ON THEIR ORIGIN. TO ARISTOTLE, THESE STRANGE BODIES WERE ATMOSPHERIC PHENOMENA. THIS IS COMET HALE-BOPP, PHOTOGRAPHED ON 29 MARCH 1997, WHEN ITS DISTANCE FROM THE EARTH WAS 196 MILLION KILOMETRES.

Anyone who ever raises their eyes to the vast expanse of the night sky begins to wonder about the nature and origin of the universe. Even today, the spectacle of the Moon's regular round, the seasonal cycle of the constellations, and the slow, majestic rise of the Milky Way give rise to the same sense of awe, and the same questions, as those they inspired at the dawn of civilization, when *Homo sapiens* first began to try to make sense of the world. Although the eternal, unchanging ballet of the Moon and the planets reflected the reassuring order and unity of the cosmos, the utterly unforeseen occurrence of a total eclipse, the passage of a comet, or the unexpected appearance of a star, were all disturbing events, like some disquieting echo of primordial chaos or a premonition of future catastrophe.

The temptation to try to comprehend the whole of the universe is probably as old as humanity. Each era had its own cosmological world-view, which reflected a greater or lesser understanding of the sky, and inherited features from a particular philosophical and mythological culture, itself more or less based on religious doctrine. The question of the macrocosm – the universe as a whole – was simultaneously both fascinating and terrifying. Yet the basic metaphysical questions have not changed greatly throughout history. Indeed, when it comes to the question of the origin of the universe, do we really have any better answer than the inspired prehistoric artists who worked at Lascaux or Combe-d'Arc?

The first questions that were truly cosmological in nature – What is the true nature of matter? Are space and time really infinite? – probably date from the time of Greek civilization. Democritus already envisaged matter as being an aggregate of atoms; Zeno ventured into the disturbing abyss opened up by the infinitely small; Epicurus wondered about the multiplicity of worlds; Aristarchus realised even in his time – and despite it being contrary to intuition and everyday experience – that the Earth revolved around the Sun. It was, however, Aristotle (384–322 BC) whose work had the most enduring influence on Western thought until the end of the Middle Ages. Unable to account for the motion of the Earth – after all, you can't feel it moving – he abandoned a heliocentric view in favour of a geocentric one. As founder of the Peripatetic School, he took the apparently spherical nature of the celestial sphere as a mirror of reality, and envisaged the universe as a series of concentric crystalline spheres, which carried the planets, and which were surrounded by an ultimate sphere, that of the so-called 'fixed' stars.

■ THE MOON AND VENUS IN THE SUNSET SKY. TO ARISTOTLE, THE HEAVENS WERE A SERIES OF CONCENTRIC, CRYSTALLINE SPHERES THAT CARRIED THE SUN, MOON, THE PLANETS, AND THE STARS. AS EARLY AS THE 3RD CENTURY BC, ARISTARCHUS REALIZED THAT THE EARTH AND THE PLANETS ORBITED THE SUN. BUT THE HELIOCENTRIC MODEL OF THE UNIVERSE HAD TO AWAIT THE RENAISSANCE AND NICHOLAS COPERNICUS FOR IT TO BECOME GENERALLY ACCEPTED.

Even before the Renaissance, some slightly earlier philosophers, such as Nicholas of Cusa (1401–1464) rejected scholasticism – Aristotle's teachings rewritten and reinterpreted by Christian theology – and wondered about the nature and limits of the cosmos. Breaking away from the concept of an enclosed universe, Giordano Bruno (1548–1600) put forward the dizzying prospect of a boundless universe, populated by an infinite number of worlds. This conception was so radically new that it – or rather its heretical connotations – caused him to be burnt at the stake on 19 February 1600 by the Holy Inquisition.

The birth of scientific cosmology

The idea of an infinite universe took root, however, in the free spirits of the Renaissance. Before they were able to define or support this theory the philosophers needed to reject the idea that the Earth was the centre of the Universe. This task fell to Nicholas Copernicus (1473–1543) and to Galileo Galilei (1564–1642). The former published his *De revolutionibus orbium coelestium* at Nuremberg in 1543, which set the Sun at the centre of the universe and, taking account of the apparent lack of motion of the stars, assigned truly astronomical dimensions to the universe. As for Galileo, who was the first physicist in the modern sense of the term and a true harbinger of relativity, he maintained that the Earth actually moved, gave his support to the heliocentric theory, and, by turning a telescope towards the night sky for the first time in 1609, opened a new era in the history of science.

By revealing thousands of new stars, Galileo's observations proved that the celestial sphere was nothing more than an illusion, and that stars really did spread throughout three-dimensional space. The Florentine astronomer also understood that the Moon and Venus, far from being the abstract geometrical concepts envisaged by Aristotle, were in fact other individual worlds. As a final, decisive factor, he recognized the ballet performed by Jupiter and its four satellites as being a Solar System in miniature.

There was nothing in this Renaissance astronomy, however, that was capable of sustaining a true scientific cosmology. It was the English physicist Isaac Newton (1642–1727) who was the first to try to generalize the laws of nature. In 1687 he published

his theory of universal gravitation: 'Two bodies attract one another with a force whose strength is directly proportional to the product of their masses and inversely proportional to the square of their distance.' The mass of the Earth is enormous relative to the mass of an apple, so it pulls the latter towards its surface. The reason why the Moon does not fall down towards the Earth is simply because the Earth's attractive force is exactly balanced by the centrifugal component of the Moon's orbital velocity, and it is this equilibrium that holds our satellite in a quasi-eternal orbit around the Earth. Newton's equations are remarkably relevant: his theory of gravitation – which he intuitively envisaged as applying to the whole universe – enables us to account for the motions of all the stars. Newton also revealed the deep unity of the universe, that is, the identical nature of apples and moons, of the Earth and the stars, and of the laws that govern them.

For the first time, the theory of universal gravitation gave scholars and philosophers a framework for cosmological thought. For example, Newton's law allowed rational questions to be asked about the infinite extent of the universe. The universe, it was realized in the 17th century, ought to be infinite and uniformly populated by stars, so that on a large scale, the attractive forces of all the bodies would cancel out.

Yet this infinite number of bodies implies a consid-erable enigma, because a simple glance at the starry sky becomes an observation of immense cosmological significance: the sky is dark at night. At the end of the 17th century, the astronomers Halley, de Chéseaux, and Olbers, all noted that an infinite number of stars, whatever their distance, would completely cover the celestial sphere, filling it with blinding light. Once the Sun has set, thousands of stars may be seen with the naked eye, and hundreds of millions of others with a telescope, yet the sky remains dark... as if space were finite and the number of stars that it contains were correspondingly restricted. This enigma, known as the 'night-sky paradox' (or 'Olbers' Paradox') remained unanswered for more than two centuries. The law of universal gravitation, which is such a powerful tool of celestial mechanics, was of no assistance at all to astronomers and philosophers in understanding this most elementary observational fact, which actually conceals some

■ In the 5th century BC Anaxagoras taught that the Sun was an incandescent rock 'larger than the Peleponnese'. In this photograph, taken on 6 November 1993, the planet Mercury is seen passing in front of the Sun (lower right, close to the solar limb).

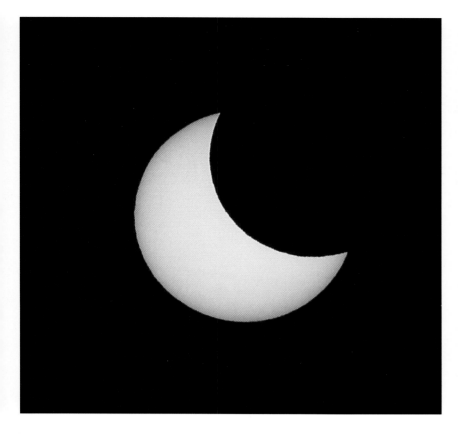

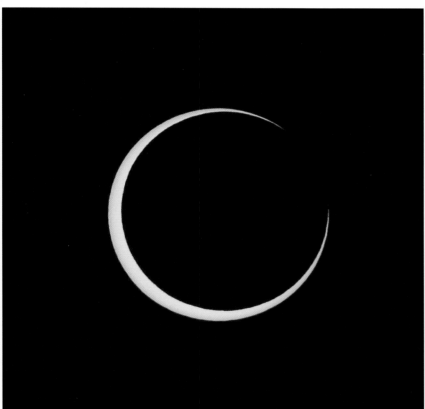

of the universe's most arcane truths.

It was not until the

ECLIPSES SATISFIED THE GREEKS IDEALS OF GEOMETRIC PERFECTION. THEY ALSO, TWENTY-THREE CENTURIES LATER, INSPIRED EINSTEIN TO DEVELOP HIS THEORY OF RELATIVITY. TO EINSTEIN, THE FORCE OF GRAVITATION WAS BASICALLY A GEOMETRICAL PHENOMENON: THE EFFECT OF THE CURVATURE OF SPACE.

Renaissance and Galileo's first essays in experimental physics, this method has proved not only

beginning of the 20th century that the fundamental questions raised by the Greek philosophers or the Renaissance scholars were finally freed from the sphere of metaphysics – which was where they had been confined for lack of a mathematical theory in which they could be expressed – and were established on an equal footing in the sphere of scientific enquiry. In the 20th century, theoretical cosmology and observational cosmology developed in parallel. As currently understood, cosmology is the science whose field of study is the universe considered as a whole. This concept, which is perhaps utopian, is based on the hypothesis (which is essentially identical with the scientific method), that the laws of physics apply universally and that the universal constants do not vary. Since the

to be the most satisfying from a philosophical point of view, but also – and most importantly – the most productive.

The world around us, even in its most distant and exotic recesses, such as the heart of stars or the interplanetary void, does in fact appear to be governed by the same set of laws of nature. Wherever we look, the same causes produce the same effects. The identification of major natural laws, which we have partly achieved, and their practical application to technology, does not, of course, imply that their true nature is understood – in so far as they are understandable. Science, which is generally extremely articulate when asked for a comment, is mute when the question is concerned with why. By its very nature, cosmology is a science

■ Beyond the sphere of the Moon, Aristotle's universe was eternal, immutable, and perfect. Although our sub-lunar world consisted of four elements, earth, water, air, and fire, the celestial bodies consisted of a fifth element, ether, which was the quintessence of purity. In 1609, when Galileo turned his telescope onto the Moon, however, he discovered another world, with other landscapes, and other individual places.

quite unlike all the others. The universe's completely unique status as a phenomenon without a cause may even mean that it falls outside the sphere of science in the strict sense. In fact, multiple keys are required to unlock cosmology (the only discipline that aims to understand 'everything'), and they are often rejected en bloc, for philosophical or religious reasons.

The framework of astrophysical theories is based on the cosmological principle, which postulates that the universe has the same properties everywhere, and that our portion of the cosmos is representative of the universe as a whole. It was tempting to generalize even further, by extending this concept to the composition of the universe over time. This was the goal of the perfect cosmological principle: the universe is the same everywhere, and at all times.

■ THROUGH OBSERVING TOTAL ECLIPSES OF THE MOON, THE GREEKS REALIZED THAT THE EARTH IS ROUND, AS SHOWN BY ITS SHADOW PROJECTED ONTO THE MOON. IT WAS BY MEASURING THE TIME THAT THE MOON TOOK TO PASS THROUGH THE EARTH'S SHADOW THAT ARISTARCHUS WAS THE FIRST TO MEASURE ITS DISTANCE RELATIVE TO OUR PLANET.

THE DISCOVERY OF RELATIVITY

For many, this concept is both extremely intuitive and philosophically ideal. Nature is, however, indifferent to our prejudices, and it was by discovering, in an extraordinary and unexpected way, that the universe does not appear to adhere to the perfect cosmological principle that the science of 'everything' really established its pedigree. Our conception of the universe changed radically immediately after the First World War. This was initiated by the commissioning of giant telescopes, which finally revealed the large-scale structure of the universe. At Mount Wilson in California, Edwin Hubble identified the true nature of the millions of faint nebulosities that people had started to discover in the sky. These, the galaxies, were vast collections of hundreds of billions of stars. The sparkling band of the Milky Way, which, according to astronomers in the 19th century comprised the whole universe, was shown to be just one of these innumerable galaxies; the one in which the Sun happened to be located.

The development of general relativity was to cause an even more radical upheaval in our concept of the universe, because it was the first mathematical theory to be teeming with innumerable different cosmologies. After 1915, the date Albert Einstein's theory was published, the 20th century saw a constant litany of successes for relativity, with an uninterrupted dialogue between theoreticians and observers, and the discovery by astronomers of essentially all of the relativistic effects predicted by Einstein. The history of cosmology is, in effect, primarily the history of relativity.

The theory of the expansion of the universe – popularly known

by the ironical term 'the Big Bang' originally suggested by its most ferocious opponent, Fred Hoyle – arose in the 1920s from this convergence of the theory of relativity with the observational discoveries made by various astronomers. On the theoretical side, the physicists explored the cosmological possibilities offered by the revolutionary theory of gravitation. The bodies in the universe – the planets, stars and galaxies – and their dynamics reveal that gravity governs the large-scale structure of the universe. Researchers therefore proposed that it also governed the behaviour of the universe as a whole. Before Einstein, ever since 1687, astronomers had been satisfied with Isaac Newton's theory of universal gravitation. The latter, which had been circulating for more than three centuries, is, in any case, a natural concept for most of us, and the obvious framework for any astronomical discussion. This is a status that the disconcerting theory of relativity is still far from acquiring. According to Newton's theory, however, the actors are quite independent of the scene that they are playing. Newtonian time and space are fixed, absolute, external frames of reference. Time passes in a linear fashion in infinite space. Despite being phenomenally successful in modelling and explaining the interactions between astronomical bodies, Newton's theory never enabled a truly scientific cosmology to emerge.

Faced with this impasse, Albert Einstein radically transformed our conception of the world. His theory of general relativity first introduced space and time as actors on the physical stage. To Einstein, time and space were no longer independent frameworks that were fixed and absolute, and within which natural pheno-mena took place. They were instead integral parts of those very phenomena. Rather than the linear time and fixed space that form our everyday experience, Einstein proposed a space-time that has variable geometry. The specific values of this space-time may vary, but wherever an observer is located, the speed of light, c, remains constant. As a result, measurements of space and time alter when two observers are situated in different reference frames. The factor that distinguishes between these reference frames, in general relativity, is their relative velocity: space contracts and time dilates as the velocity increases.

In everyday life, where speeds are ridiculously small in comparison with c, these strange effects (known as relativistic effects) are imperceptible. But imagine twin sisters, one living on Earth, and the other in a spaceship that is moving at a speed close

to that of light. Each sister will undergo, in her own frame of reference, the same sort of experiences and any physical measurements that they make (of the speed of light, for example) will give identical results: this is the principle of invariance. On returning to Earth, however, the intrepid astronaut will find that although just a few months of her own time have elapsed, her sister will be several years older. Because the speed of light remains a constant (and speed is distance divided by time), time, on board the rocket, dilated – i.e. it passed more slowly than on Earth. Lacking rockets that can move at the speed of light, physicists have confirmed the dilation of time by using two atomic clocks, one on board an aircraft, and the other remaining on the ground. In accordance with Einstein's equations, the clock on board the aircraft gained an infinitesimal amount on the clock that remained on the ground. The constant nature of c, and also its absolute nature – nothing can move faster than the speed of light in a vacuum, 299 792 458 km/s – give it the status of a magic number: the speed of light is the foundation of the theory of relativity.

The light-year is the simplest and the most aesthetically pleasing unit for measuring space-time. It is the distance covered by light in one year: in practical terms 10 000 billion kilometres. Because light is the fastest messenger in the universe, any information from any point whatsoever in the universe propagates at that speed. The 'light-distance' is not only a convenient astronomical unit, but also an extremely evocative one, because it enables us to understand the space-time nature of the universe. So, for example, the farther we look out in space, the farther we look back in time. Within the Solar System, of course, this time delay is very restricted: only about one-and-a-half seconds for the Moon, for example. When it comes to the Sun, the time delay is rather greater. When we admire the setting Sun at the seaside, we are actually seeing it as it was over eight minutes ago. The closest star to the Sun, Proxima Centauri, lies at a distance of more than 40 000 billion kilometres, i.e. at 4.2 light-years. Deneb, in Cygnus, one of the most distant stars visible to the naked eye, is more than 3000 light-years away.

Einstein's equations do not just establish the speed of light as a constant. They also enable us to obtain a completely new point of view regarding a fundamental feature of nature that no one notices, because it is so firmly rooted in our everyday experience, but which nevertheless has intrigued scholars and philosophers since the Renaissance. This is the principle of equivalence, that is the uniform behaviour of bodies to the 'force of attraction'. Countless real or thought experiments have shown that bodies that consist of different materials and of different masses behave exactly the same under the force of gravity. Whether we are dealing with the objects that Galileo (allegedly) dropped from the Leaning Tower of Pisa, or with those carried into space by astronauts – in free fall, or weightless, respectively – their behaviour is identical. Why? In answering this question, Einstein abandoned Newton's 'force' of gravity. To him, gravity is a geometrical characteristic of space-time. It is its curvature. How can we accept a concept that is so abstract, so strange, and apparently so unnatural? Relativity requires a radical change in our ways of thought. Eighty years after publication of the theory, it still does not form part of our common cultural heritage. Nevertheless, although the theory of relativity is a complex intellectual creation, which is completely abstruse to non-scientists, at odds with common-sense, and, above all, apparently unrelated to the real world, its effects, however strange and exotic they may seem, are undeniably observable nowadays. In fact, all the predictions made by Einstein and the relativists that followed him have either been observed in nature or reproduced in the laboratory.

Relativity's success does not, on the other hand, imply either that the theory is complete or that it is definitive. Remember that Newton's view, despite being so profoundly different from Einstein's, is still perfectly adequate for explaining the behaviour of celestial bodies. What will be the status of relativity in the future? Will it withstand three centuries of astronomical observation, as did its predecessor? Will it be replaced by a more general theory that will incorporate its features, or will it be swept away by some new paradigm, some blinding flash of insight by some future genius? Without trying to second-guess the future, it currently seems unlikely that any future theory will dispense with the fundamental concept that has been sustained by the theory of relativity, namely the curvature of space-time.

The Big Bang revolution

It has been the exploration of the theory of general relativity as much as actual observation of the sky that has prompted the rapid development of cosmology in recent years. Its equations contain the seeds for an infinite number of universes, and for eight decades scientists have been seeking to discover in which of them we live. It is easier to comprehend the extraordinary richness of general relativity if we recall that the first exact cosmological solutions to Einstein's equations were found even before astronomers had the slightest inkling of the actual architecture of the universe.

In 1917, the father of relativity himself published an initial model of the universe, which obeyed the perfect cosmological principle. Einstein's universe is static, isotropic, and exists for an infinite time. In those days no one had any clear image of what space was really like beyond the Solar System. Was the Milky Way the whole universe, or was it just one among millions of 'extragalactic nebulae'? To obtain a model of the universe that was credible at that particular time, in other words, a static universe, Einstein introduced into his equations a term, known as the cosmological constant, whose validity, existence, and hypothetical value are still being discussed today. This was, as he said later, 'the greatest error of my life'. This *ad hoc* constant was designed to stabilize a universe which, according to his equations, Einstein had found to be inherently unstable.

In 1922, when the nature of the galaxies was still being debated, the Russian, Aleksandr Friedmann discovered in general relativity strange cosmological solutions, in which the universe spontaneously expands. Finally, in 1927, a Belgian mathematician, Georges Lemaître, published the first modern cosmology. Beginning with the selfsame equations of general relativity, Lemaître postulated a universe that expanded from an initial explosion, which would have taken place billions of years ago. For the first time, he supported his theory by pointing to astronomical observations, which, even then, tended to show that the 'extragalactic nebulae' were receding from the Milky Way. It was in the same year of 1927, in fact, that the American astronomer Edwin Hubble caused a sensation by publishing his famous law that all galaxies appear to be receding from one another at a velocity that increases in proportion to their distance.

Although supported by observation, the theory of an initial explosion nevertheless encountered a lot of reservations, not least for philosophical reasons. Triumphant 20th-century material-ism did not really sit well with any violation of the perfect cosmological principle. An

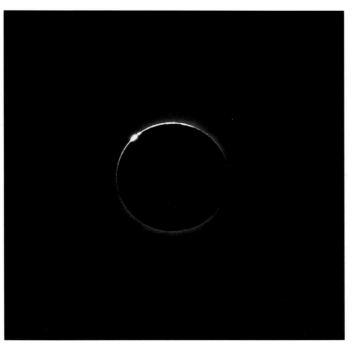

■ In his theory of general relativity, Einstein proposed for the first time that space is curved in the presence of mass. In 1919, during a total eclipse, observations confirmed the effect of this curvature on the positions of stars close to the Sun.

annoying factor was the fact that because Georges Lemaître was a priest, his most implacable opponents wrongly saw his suggestion of an initial explosion as a allegory for the *Fiat lux!* of the Bible. Following on from the discovery of the expansion of the universe, however, the initially shaky edifice of the Big Bang was to be reinforced by two other concrete pieces of evidence. Analysis of the composition of the stars was to show that they do not exist for all eternity, but that they evolve, and that the universe thus has a long history behind it. Second, in 1965, the cosmic background radiation, a sort of relic of a hotter and denser stage of the universe, was discovered accidentally by Arno Penzias and Robert Wilson. The theory of an initial explosion has enjoyed great success ever since. Although the mathematical and physical concepts have changed greatly over the years, it is possible to say that the present-day canonical cosmological model still remains that of Georges Lemaître, together with its conceptual framework, general relativity.

It is the Big Bang model that has enabled us to resolve the mysterious paradox of why the sky is dark at night. According to the models of the universe that Einstein has bequeathed to us, the light that we receive from the stars does not travel at an infinite velocity. Our vision of the world is, in fact, limited by a fundamental horizon, which corresponds to the distance that light has covered since what we take to be the origin of the universe. So, whether the universe has a limit, or is infinite, the celestial sphere that we admire at night is not without a boundary. The number of stars that lie within this apparent shell surrounding us, and which is expanding at the speed

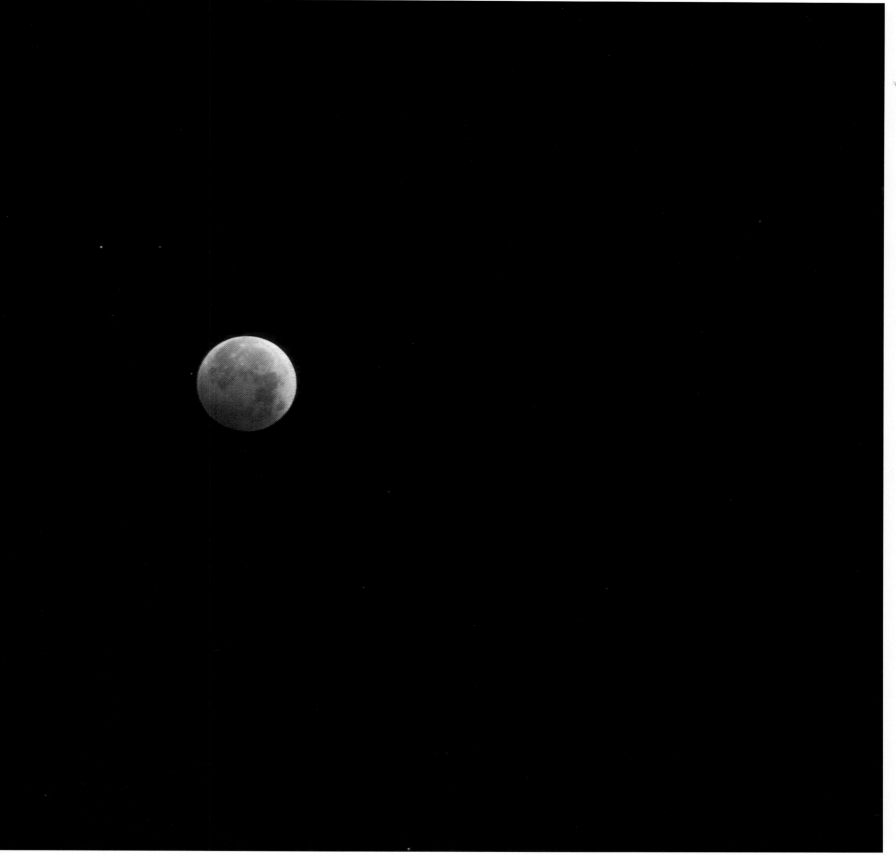

of light, is also limited. The some 100 000 billion stars that riddle our region of the universe are insufficient to drive away the darkness.

Modern cosmological theories are fruitful because they may be refuted; no sooner are they conceived than they are confronted by the observational data obtained by large telescopes. The results leave very little doubt about the validity of the expansion theory; they enable us to visualize the curvature of space, to demonstrate the recession of the galaxies and the evolution of stars, and are now capable of pushing back in time to the fringes of the Big Bang itself.

Any hope of understanding the universe as a single mental image will always remain an unattainable dream, even though some physicists believe that they will one day establish a 'theory of everything', capable of explaining even the origin of the universe itself. Faced with this dizzying fractal spiral, the two incomprehensible ends of which appear to be lost, one in the infinitely large, and the other in the infinitely small, we shall probably remain with our heads in a whirl for a long time to come, and the very first questions humans posed about the nature of the universe will remain without answers. ■

The Galaxy: an island in space

■ THE MILKY WAY APPEARS TO ENCIRCLE THE CELESTIAL SPHERE. IT IS, IN FACT, THE GALAXY, IN WHICH THE SUN IS LOST AMONG HUNDREDS OF BILLIONS OF STARS. ON THE LEFT, WITHIN THE CONSTELLATIONS OF SAGITTARIUS, SCORPIUS, AND OPHIUCHUS IS THE GALACTIC BULGE, THE DENSEST AND BROADEST PART OF THE MILKY WAY. THE TWO MAGELLANIC CLOUDS, SMALL SATELLITE GALAXIES OF THE MILKY WAY APPEAR JUST ABOVE THE HORIZON.

■ THE GALACTIC CENTRE, WHICH LIES ABOUT 26 000 LIGHT-YEARS AWAY, IS CONCEALED
BY DENSE CLOUDS OF INTERSTELLAR GAS AND DUST THAT HAVE ACCUMULATED IN THE
DISK OF THE MILKY WAY.

At the height of northern summer, the finest skies may be seen from the high mountains, far from city lights and above the clouds and mists of the lower atmosphere. From those peaks where the world's great observatories cluster, and ramblers and climbers camp in search of altitude and solitude, the bright stars of Scorpius rise above the south-eastern horizon. Ursa Major, the Great Bear, in the course of its eternal circling of the unmoving Pole Star, grazes the rugged silhouette of the northern horizon.

In the dark sky dotted with stars, the Milky Way begins to appear once the last light in the west has faded. Initially faint and misty and hard to see, it soon makes itself felt as the night darkens. Like some vast embroidered scarf, it stretches from one horizon to the other, from Sagittarius to Perseus, crossing the entire sky, and lighting up the constellations of Aquila, Cygnus, and Lyra. Far to the south, on the other side of the Equator, the Milky Way is resplendent during the southern winter, being accompanied by some of the most beautiful stars in the sky: Sirius, Rigel, Canopus, Achernar, the twin stars of Centaurus, and the scintillating diamonds of the Southern Cross.

The Milky Way thus appears to encircle the whole of the terrestrial sky. In the absence of any reference points, the human brain projects all the stars onto a purely imaginary celestial sphere, but one that appears so tangible that the ancients believed that it really existed. The sparkling arc thrown across the stars should, in fact, be seen as our Galaxy, an enormous disk of stars viewed obliquely, and within which we are actually located. The Milky Way represents a section of this disk, seen from the side. The number of stars within it is so great, and the distances at which we can see them are so large that they merge into a sparkling mist. Faced with the spectacle of this faint light arching across the whole sky, it is nevertheless possible to imagine ourselves outside it, and see it as a disk of stars. It is, on the other hand, quite impossible to form any idea of its immense size. The Galaxy could well be the whole universe, and indeed until the beginning of the 20th century, astronomers thought that it was just that. That was before the discovery of other galaxies, an infinite number of similar islands in space that populate the depths of the universe as far as astronomers can see.

The Galaxy is a rotating disk of stars, swollen at its centre. It

■ THE SPIRAL DISK OF THE MILKY WAY MAINLY CONSISTS OF STARS. IN THE GALACTIC PLANE, WHICH CORRESPONDS TO THE DENSEST AND YOUNGEST REGIONS, GASEOUS NEBULAE HAVE FORMED, SUCH AS M 17 (BOTTOM), IN THE CONSTELLATION OF SAGITTARIUS. ITS REDDISH COLORATION IS THE SIGN OF THE HYDROGEN GAS OF WHICH IT IS COMPOSED, HEATED BY THE LIGHT FROM NEIGHBOURING STARS.

■ 'BAADE'S WINDOW' IS A REGION
OF THE MILKY WAY THAT IS
RELATIVELY FREE OF INTERSTELLAR
DUST WHERE ASTRONOMERS ARE
ABLE TO OBSERVE SOME MILLIONS
OF THE BILLIONS OF STARS THAT
POPULATE THE GALACTIC BULGE.
MOST OF THEM LIE MORE THAN
10 000 LIGHT-YEARS AWAY
FROM EARTH. THE BRIGHT
STAR GAMMA SAGITTARII,
(TOP LEFT), IS JUST 96
LIGHT-YEARS AWAY.

■ When photographed with a large telescope, the Milky Way appears like some dense stellar fog. The appearance is deceptive: several light-years, that is to say, some tens of thousands of billions of kilometres separate each star from its nearest neighbour.

measures more than 80 000 light-years in diameter, 16 000 light-years thick over the central bulge, and about 1200 light-years thick in the vicinity of the Sun, which lies towards the outer edge of the disk, at 26 000 light-years from the galactic nucleus. The nucleus itself is undoubtedly an extremely massive body, but its nature is still unknown. These are the visible parts of the Galaxy, but the galactic disk is surrounded by an invisible spherical shell, the halo, the size of which is uncertain. It is certainly more than 100 000 light-years in diameter. Within this halo orbit more than 100 globular clusters, all approximately spherical, and each one of which consists of hundreds of thousands of stars, closely packed together.

The geography of the Milky Way

The incredible density of the Milky Way, where the stars appear to be almost literally crammed together, is an illusion caused by the vast distances involved. In reality, the Galaxy is essentially empty. The average distance between stars is actually about one hundred million times their diameters. This is as good as saying that in their orbits around the galactic centre, the tens or hundreds of billions of stars in the Milky Way run no risk of crashing into one another. The closest star to the Sun, Proxima Centauri, lies at a distance of 4.2 light-years, slightly more than 40 000 billion kilometres away. Next come the pair, Alpha Centauri A and B, at 4.4 light-years, Barnard's Star at 5.9 light-years, and then Wolf 359, at 7.8 light-years. Apart from them, there is nothing except empty interstellar space. To get a better grasp of these distances, and in particular, of interstellar space, imagine that the Sun has been reduced in size to that of an orange, which we place in the centre of Paris. On this scale, Proxima Centauri – a small clementine – would lie 4000 km away, in St Johns, Newfoundland; Alpha Centauri A and B, a pair of oranges, would be about 4200 km away, between Newfoundland and Nova Scotia; Barnard's Star, another clementine, would be about 6000 km away, in New York; and Wolf 359, the size of a plum, would be 7500 km away in New Delhi. Even at this reduced scale, the entire Galaxy would be an immense disk, populated by tens of billions of grapefruit, oranges, and clementines, some 100 million kilometres in diameter!

From our terrestrial viewpoint – with its false sense of perspective – the overall shape of the Galaxy is nevertheless recognizable when we scan the Milky Way from one horizon to the other. A narrow section of the disk crosses the whole sky. In the constellation of Sagittarius, which lies in the direction of the galactic centre, the thickness and the brightness of the Milky Way increase significantly, indicating the presence of the central bulge. Other galaxies, when seen side-on through a telescope, have the same profile. It is by comparing the images of thousands of galaxies, seen from all angles, that astronomers have been able to create an Identikit picture of our own Galaxy. We shall probably never be able to see the Galaxy from outside; the distance that we would have to cover to see it from a distance is never likely to be attainable. At the speed of one of the fastest of modern-day rockets, it would take some ten million years for some intrepid – and exceptionally patient – adventurer to get far enough away from the galactic disk to be able to see it in its entirety.

Like most astronomical bodies, the Galaxy is rotating, carrying its horde of stars around the galactic centre. In the region of the Sun, stars are moving at about one million kilometres an hour. At this speed, it takes the Sun at least 250 million years to complete one orbit of the Galaxy. As it covers the light-years, it passes through regions of the Galaxy that are richer or poorer in stars, encountering millions of other stars, all on different orbits. Slowly, the sky as seen from Earth changes. The constellations, which seem to us to be immutable and eternal, have not always existed in their current form. Imperceptibly, their shapes are changing. At the dawn of civilization, people saw different patterns in the sky and, one day, the Great Bear and the Southern Cross will have disappeared, borne away by the stream of time. Since it was formed, about 4.6 billion years ago, the Sun has completed twenty-odd galactic orbits, but there was no one on Earth to marvel at the spectacular objects that it has probably approached from time to time. These may have included flamboyant nebulae like vast auroral curtains; blue supergiants that illuminate the night with a flickering light, casting ghostly shadows across the landscape; supernovae too bright to look at, showering down fatal invisible radiation; fiery Cepheids pulsating with a throbbing beat... But the life of the Galaxy does not flow on the same scale as those of

the stars, nor like that of the civilization that one of them has nurtured. When the Sun started its last galactic orbit, 250 million years ago, the Earth was home to giant insects and reptiles. The latter were about to give rise to the dinosaurs. The orbit before that, the continents were barren, and only the sea was populated with arthropods. One orbit earlier, and the jellyfish, the corals, and the innumerable unicellular organisms that were already teeming in the water, were all undoubtedly indifferent to the wonders in the heavens.

The galactic population

Although the dimensions of the Galaxy – except for those of the mysterious halo – are nowadays well-known, that is to say, with uncertainties of a few hundred light-years, no one knows how many stars are to be found in the vast disk of the Galaxy. With the naked eye, it is possible to count, on a clear night, more than three thousand. This includes just a few very bright stars that are close neighbours. Most of the stars visible to the naked eye lie within a sphere of just 100 light-years in diameter. In the dense clouds of stars in the Milky Way, where astronomers can in places reach out to more than 20 000 light-years, telescopes are able to record several billion stars. In reality, however, the galactic population far surpasses that figure.

It is possible to make a crude estimate of the number of stars in the Milky Way, simply by comparing the total luminosity of the Milky Way with that of the Sun, regarded as a typical star. This estimate gives a galactic population of more than 100 billion stars. Is there any way of comprehending such a vast figure? If anyone wanted to catalogue all the stars in the Galaxy, counting them one by one at the rate of one a second, it would take more than 3000 years to complete such an impossible survey.

In any case, it is at present quite impossible for us to see all of these stars. On the one hand, the smallest, which are extremely numerous, are very faint, perhaps only one-hundred-thousandth that of our own star. None of these dwarf stars is visible to the naked eye, and even the most powerful telescopes are unable to detect them at distances greater than about 1000 light-years. On the other hand, the galactic disk is not transparent: substantial clouds of gas and of interstellar dust, produced by the reactions occurring in

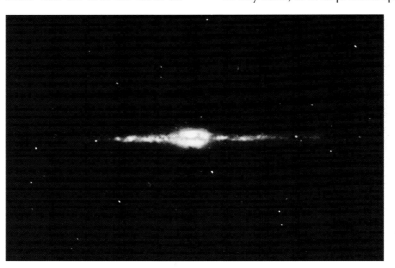

■ THE MILKY WAY, PHOTOGRAPHED IN INFRARED LIGHT BY THE AMERICAN COBE SATELLITE. THIS WIDE-FIELD IMAGE, WHICH WAS OBTAINED IN 1992 AND RECREATED BY COMPUTER FROM SCANS OF THE COMPLETE CELESTIAL SPHERE, SHOWS PERFECTLY THE GALATIC DISK AND CENTRAL BULGE.

stars, absorb light from objects that lie behind them. In places, the Milky Way is even completely opaque. To the naked eye, these areas of sky appear extremely dark, including the famous region in the southern Milky Way known as the 'Coalsack'. The figure of 100 billion stars therefore corresponds to the number of stars that astronomers would be able to see in the Galaxy, if their telescopes were powerful enough, and if interstellar dust clouds did not hide them.

THE GALAXY'S OVERALL MASS

Despite being so large, this figure does not actually take account of the total mass of our Galaxy, which may be estimated by a more radical method than simple observation of the stars. This involves measuring its overall gravitational field. This method has the great advantage of including all astronomical objects, whether they are visible with our telescopes or not. According to the law of gravitation, the greater the mass of the attracting body, the faster an orbiting body moves around it. In addition, the attractive force decreases with distance. In the Solar System, for example, the closest planets, subject to the powerful gravitational pull of the central mass of the Sun, need to orbit very rapidly to counteract the otherwise fatal attraction. The distant planets, in contrast, are less subject to the Sun's gravitational field, and orbit at less than one tenth or one hundredth the rate.

Astronomers expected to see the same sort of behaviour among stars in the Galaxy. There was nothing of the sort. In fact, they were amazed to find that, on the contrary, as the distance of stars from the galactic centre increased, so did their velocities! This phenomenon, which has been confirmed by numerous measurements over several decades, has only one explanation: the overall mass of the Galaxy is actually far greater than we perceived. The visible stars merely amount to the tip of the galactic iceberg. It is currently estimated that the mass of the Galaxy – taking an arbitrary diameter of just 100 000 light-years – is at least 200 billion solar masses. But the most recent observations tend to show that the galactic halo extends much farther than we previously suspected, and might measure as much as 500 000 (or even one million) light-years across. The total mass of the Galaxy might then be as much as 1000 billion solar masses. How can we explain the fact that just 10% of the mass of our Galaxy is observable?

Astronomers don't know the answer. They have tried every way of detecting this 'missing mass' (also known as 'dark matter'). This mystery – the identity of this formidable amount of material of an unknown nature – has become the greatest problem in astrophysics at the end of this century (see Chapter 11). In fact, just like the Milky Way, all the galaxies have a

■ This group of stars lies on the borders of the constellations of Scorpius and Ophiuchus. Opaque clouds of dust hide background stars. A few supergiants that were born recently illuminate the nearby veils of dust. These nebulae appear bluish, because they scatter light from the stars, similar to the way in which the Earth's atmosphere scatters light from the Sun.

gravitational mass greater than their visible mass. The only thing we can be sure of is that the missing mass does not consist of stars – we would see them. Not only is this material invisible to our telescopes, it is also invisible at all wavelengths of the electromagnetic spectrum. It makes its presence felt solely because of its mass. We do now know that this invisible matter appears to be distributed even throughout the galactic halo. Its considerable mass pulls the disk of the Milky Way along with it.

■ THE MILKY WAY IN THE CONSTELLATION OF MUSCA. A VAST CLOUD OF DARK INTERSTELLAR DUST HIDES THE STARS THAT LIE BEHIND IT. THE GLOBULAR CLUSTER NGC 4372, A SPHERICAL CONDENSATION OF SEVERAL HUNDREDS OF THOUSANDS OF STARS THAT LIES 15 000 LIGHT-YEARS AWAY, IS SLOWLY CROSSING THE GALACTIC PLANE.

Strangely, the probable existence around us of an invisible mass amounting to 1000 billion solar masses does not appear to have any influence on the general processes that occur within the visible portions of the Galaxy. The cycle of birth, evolution, and death of stars; the mixing that occurs in the nebulae; and the chemical evolution of the various bodies are apparently completely unaffected by the presence of dark matter: to such a degree, that specialists in galactic evolution do not take it into account when they study the structure of the Milky Way in detail.

Our Galaxy is a wondrous mechanism for creating stars. Probably born some 12 billion years ago, under conditions that cosmologists are still trying to work out, it originally primarily consisted of gas – the famous primordial hydrogen left over from the Big Bang. Nowadays, 90% of its mass – its visible mass, of course – consists of stars, the remaining 10% forming the vast clouds of interstellar gas and dust known as nebulae, some of which are so large and bright that they are visible to the naked eye. Mixing that occurs within the gas continually gives rise to

■ IN THE CONSTELLATION OF SAGITTARIUS, THE LAGOON NEBULA (BOTTOM) AND THE TRIFID NEBULA (TOP) LIE AT A DISTANCE OF ABOUT 5000 LIGHT-YEARS. THERE ARE SEVERAL DOZEN, EQUALLY BRIGHT NEBULAE IN THE GALACTIC DISK. THEIR MASS AMOUNTS TO SEVERAL THOUSAND SOLAR MASSES.

new stars, and the process is promoted by the stars themselves, which – at least in the case of the most massive stars – return a large portion of the gas from which they were born when they explode at the end of their brief lives. The Milky Way is a true stellar nursery, and it is within its spiral arms that the youngest stars are found.

A SPIRAL GALAXY

Several vast spiral arms spring from the central bulge and gracefully stretch out through the disk, resembling the shape of an Earthly cyclone. These spiral arms have no true physical existence. They are the result of a density wave that propagates round the Galaxy. When this wave arrives, the interstellar gas is compressed and heated. This causes nebulae to form along the spiral arms. These, in turn, give birth to thousands of stars. The brightest of these, whose life-expectancy is just a few million years, briefly illuminate the spiral arms, and die out as the arm passes. This density wave circulates around the Galaxy in the same direction as the general population of stars, but far more slowly. Whereas most of the stars take somewhere around 200 million years to complete an orbit, the spiral wave takes about 400 million years. In fact, it is the stars and gas in the Milky Way that catch up with the wave, are slowed down as they pass through it, and then speed up once the wave has passed. The origin of the Galaxy's density wave has not been determined. Should we see it as a gravitational perturbation within the disk of stars, caused by some ancient encounter between the Milky Way and another Galaxy? The equivalent, in effect, of a water ripple, on a

galactic scale? Quite apart from its origin, astronomers are still wondering about the physical processes that allow the spiral structure to persist for billions of years.

If we were able to admire our Milky Way galaxy from a distance, say from one of the remote stars in the halo, and could speed up time until we could see the Galaxy turning before our eyes, we would see a gigantic, tentacled creature come to life. The 100 billion pinpoints of light that are the stars, flickering as a result of stellar evolution, would be carried around by the giant galactic whirlpool. Here and there are beautiful emerald bubbles, shells of gas ejected by old pulsating stars, opening like flowers for a short period before they vanish, carried away by the currents of space, while the slower spiral wave appears to flow upstream against the tide of stars. Ahead of the wave, long dense clouds, as turbulent and dark as storm-clouds, accumulate, being heaped up by the action of the wave. Along these interstellar clouds, flicker hundreds of sparks, budding stars that illuminate clouds of hydrogen. Among the blood-red colour of the

ABOUT ONE MILLION YEARS AGO, S MONOCEROTIS IS A
BLUE SUPERGIANT, MORE THAN 10 000 TIMES THE
LUMINOSTY OF THE SUN.

nebulae, the light from a supergiant occasionally appears, arising only to die an instant later, drowning out the whole of the Galaxy in a blinding flash of light. How can we expect to pick out our own Sun from this maelstrom of light?

Invisible, but omnipre-sent, the mysterious halo makes its presence felt by the motion that it imposes on the whole Galaxy. The latter appears to be held together by some sort of invisible glue, carried along almost as a single object by some hidden, and incredibly powerful hand. ■

31

A thousand generations of stars

■ Right in the Milky Way, and in the very centre of the constellation of Orion, is the brilliant star Alnitak. Easily visible to the naked eye, this blue supergiant 10 000 times as bright as the Sun, lies at a distance of 800 light-years. The fierce ultraviolet radiation from Alnitak heats the hydrogen in the nebulae around it, NGC 2024 and IC 434. The beautiful Horsehead Nebula, Barnard 33, is a cold, dark condensation of interstellar gas and dust.

THE FOUR SUPERGIANT STARS OF THE TRAPEZIUM IN ORION HEAT THE INTERSTELLAR GAS SURROUNDING THEM OUT TO DISTANCES OF SEVERAL TENS OF LIGHT-YEARS. THIS GAS PRIMARILY CONSISTS OF A MIXTURE OF HYDROGEN, HELIUM, AND OXYGEN.

f we think of the Milky Way – like the rest of the tens of billions of galaxies that cover the celestial sphere – as a cell, then stars are its atoms. Minute specks on the scale of the universe, which is essentially empty, they concentrate its matter, distribute its energy, and illuminate its various regions. But the role of stars is not confined to that of sparkling lights, punctuating the vast empty spaces here and there. It is stars that transform the universe. They are the true alchemists. Starting with primordial hydrogen and helium, the simple atoms that arose from the blistering heat of the Big Bang, they synthesize elements that are heavier and heavier, more and more complex. They create the silicon and iron from which the Earth and its sister planets were born; they exhale the oxygen and nitrogen that we breathe; and they and scatter masses of platinum, silver and gold without a second thought.

Our own star, the Sun, hidden away in the slowly rotating galactic disk between the Orion and Cygnus spiral arms, is indistinguishable from the myriad stars that surround it. The Sun's age, size, luminosity, and chemical composition make it a typical member of the galactic population. The physical processes discovered in the centre and at the surface of our own star are found everywhere. For these reasons, when astronomers discuss the properties of stars, they always compare them with those of the Sun.

Our own star is a sphere of hot gas, 1.4 million kilometres in diameter. Its mass, two billion, billion, billion tonnes (2×10^{27}t), is 1000 times as great as all the planets in the Solar System combined. The mass of the Sun is 330 000 times as large as that of our insignificant planet Earth. The Sun's gases, like those of most stars, consist of about 70% hydrogen, 28% helium, and 2% of heavy elements, such as carbon, nitrogen, oxygen, neon, etc. Because it primarily consists of very light elements, the Sun, like most stars, has a low density of just 1.4, i.e. not much greater than that of water, which, by definition, is equal to 1. The temperature of the Sun's blinding surface is around 5500°C, and the whole star emits an enormous amount of energy every second, corresponding to 390 million billion billion watts (3.9×10^{26}W). Where does such an amount of energy come from? Because of its gravitational attraction, the Sun's enormous mass naturally tends to cause it to collapse in on itself. At the centre, the matter is subject to an enormous pressure. In the central region the temperature of the gas is

■ M 42 IS UNDOUBTEDLY THE
MOST BEAUTIFUL NEBULA IN THE
SKY. ITS WISPS OF GAS ARE
ILLUMINATED, HEATED, AND SHAPED
BY THE FOUR SUPERGIANT STARS IN
THE TRAPEZIUM, WHICH WERE
BORN A FEW HUNDRED THOUSAND
YEARS AGO. THE NEBULA LIES IN
THE CONSTELLATION OF ORION,
1500 LIGHT-YEARS AWAY. IT IS A
FERTILE NURSERY FOR STARS,
AND IS PERHAPS THE MOST
PROLIFIC REGION IN THE
WHOLE GALAXY.

■ THE TRIFID NEBULA, M 20,
IN THE CONSTELLATION OF
SAGITTARIUS, IS 15 LIGHT-YEARS
ACROSS AND CONTAINS ABOUT
200 SOLAR MASSES OF GAS. THE
GAS IN THE NEBULA IS EXTREMELY
TENUOUS, BECAUSE EVEN AT ITS
HEART ONE CUBIC CENTIMETRE
CONTAINS JUST 100 ATOMS OF
HYDROGEN. BY COMPARISON,
ONE CUBIC CENTIMETRE OF
THE EARTH'S ATMOSPHERE
CONTAINS 100 BILLION
BILLION ATOMS.

around 15 million degrees. The thermal agitation of the atoms in a gaseous medium under a pressure 100 billion times the atmospheric pressure on Earth is so great that occasionally when hydrogen atoms collide they are forced to undergo fusion. Fusion of four hydrogen atoms produces one atom of helium. This simple atomic transmutation is accompanied by an enormous release of energy. During this process an infinitesimal part of the matter within the star is completely converted – via Einstein's famous equation $E = mc^2$, which establishes the intimate relationship between matter and energy – into a vast amount of radiation. The cores of stars are thermonuclear reactors.

Each second, the Sun converts four million tonnes of matter into pure energy. This fantastic amount of energy, together with the pressure of the gas, tends to cause the matter inside the Sun to expand, just as steam pushes up the lid of a kettle. But this radiation pressure is precisely counterbalanced by the gravitational attraction acting in the opposite direction. The Sun is thus held in what is known as hydrostatic equilibrium. For the Sun, as for most stars, the incredible loss of mass through nuclear fusion – 100 000 billion tonnes per year! – is still negligible by comparison with its total mass. Basically, our star can continue burning hydrogen for billions of years.

The light that the Sun showers us with does not come directly from its nuclear core. In fact, most of the particles emitted by the fusion reactions are gamma-ray photons, the most energetic sort of radiation. These photons are absorbed and re-emitted an in-calculable number of times by the atoms in the solar gas, slowly migrating towards the surface as this occurs. A photon from the core of the Sun takes about one million years to reach the surface, where it appears primarily as visible light, but also in the form of X-rays, ultraviolet, infrared, microwave, and radio emission.

THE BIRTH OF STARS INSIDE NEBULAE

Every year, about twenty new stars appear in our galaxy. Unfortunately, these events are invisible, because the stars are born unseen, hidden in the thick cocoons of the nebulae that are scattered throughout the disk of the Milky Way, and which pick out its brilliant arms. Enormous condensations of gas, the nebulae mainly consist – like the stars to which they will give birth – of hydrogen, helium, and oxygen, as well as interstellar dust. The latter is in the form of fine grains, extremely small particles of silicates and graphite, that are only about 1/10,000 mm in size. Slowly carried round by the swirling Galaxy, nebulae extend over a few tens of light-years. Some are cold, opaque, and dark, revealing their existence by ominous dark patches in the Milky Way that are devoid of stars. Others are hot and thus luminous, with marvellous, glowing diaphanous veils that spread across the sky like exotic tropical butt with purple or turquoise wings, or pale moths, shimmering like emeralds in the night.

The most beautiful of them, in the constellations of Orion, Monoceros, Ophiuchus, and Sagittarius, are illuminated by dozens of young stars that were born just a few hundred thousand years ago. The extremely energetic radiation from these stars is capable of lighting up space for tens of light-years around. When the star's radiation is intense enough to excite the gas, i.e. to strip away electrons, the gas clouds glow like the aurora borealis. Clouds of hydrogen gas are clothed in a characteristic pinkish-red glow, while oxygen emits radiation that is a ghostly green. The mass of these interstellar clouds may be enormous, often exceeding 1000 billion billion billion tonnes – more than enough to make hundreds of stars. Nevertheless, just like the space that they would otherwise fill, they are essentially empty. On average, a nebula has a few thousand atoms per cubic centimetre, at the most. By

■ THE PLEIADES ARE A CLUSTER OF STARS THAT WERE ALL BORN TOGETHER FROM A NEBULA ABOUT 70 MILLION YEARS AGO. THIS REMARKABLY RICH GROUP IS DOMINATED BY FIVE BLUE SUPERGIANTS AND INCLUDES SEVERAL HUNDRED STARS. IT LIES AT A DISTANCE OF 360 LIGHT-YEARS.

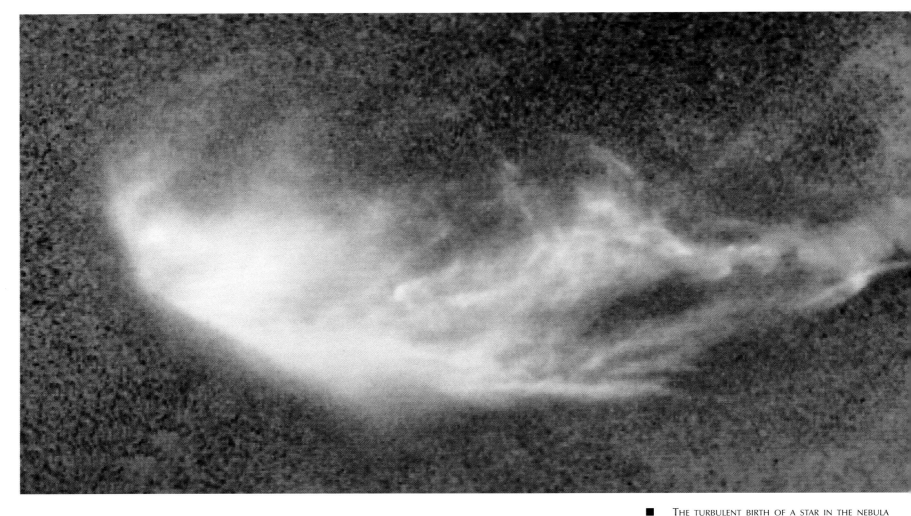

comparison, a similar volume in a cloud in the Earth's atmosphere contains more than 100 billion billion atoms.

These vast interstellar clouds are occasionally disturbed. The passage of the density wave that propagates around the galactic disk roughly every 400 million years, or the explosion of a nearby old, massive star, may cause shockwaves that compress the gas locally. The nebula may be distorted and tattered like storm-blown clouds. Here and there, where the density of the nebula increases, more massive condensations form, which tend to collapse inwards, under the force of gravitation.

These small clumps and globules of gas and dust that are scattered throughout the nebula stand out against the luminous background. They are just a few light-days across and they are completely opaque. The material within them, which is in free fall towards their centres, is so dense that no light can penetrate through them. In most cases, they can be detected by their infrared radiation, i.e. the heat that they emit. These small, slightly elongated, globules resemble eggs. They become flattened as a result of their own gravity and at the same time they start to rotate. In the very centre of these disks, an embryonic star, a protostar, forms as a sphere of denser and hotter gas. Although it has grown from the surrounding matter, it slowly begins to separate from it, blowing the material away as it starts a life of its own.

Several hundred of these Sleeping Beauties, stars hidden away in the galactic cocoons from which they are born, lie in the Great Nebula in Orion, about 1500 light-years away. Astronomers are keeping a close watch on them, because they hope that one day in the near future they will be able to capture the first rays of light from one of these stars as the veils of gas and dust around it disperse.

One of these future stars, HH 30, is even closer to us, at a distance of just 450 light-years, in the constellation of Taurus. Although it is still invisible, it has made its presence known in a spectacular manner, by ejecting into space part of the disk from which it is being born. In 1994, using the Hubble Space Telescope, for the first time astronomers were able to observe in detail the disk of matter that surrounds this incipient star, and the jet of hot gas that it has ejected at a very high velocity. The researchers even succeeded – by photographing HH 30 at yearly intervals – to detect the movement of the perfectly straight jet, which is about one light-week long, and which is emerging at a velocity of about 800 000 km/h from the developing star. In a protostellar disk, the material that falls onto the protostar increases its mass and temperature, which eventually leads to the onset of nuclear reactions, an event which marks its birth as a true star. At the same time, the body – perhaps because of the effects of an intense magnetic field – expels some of its mass at the two poles.

■ AS SEEN BY THE HUBBLE SPACE TELESCOPE, THE DISK OF GAS AND DUST SURROUNDING THE YOUNG STAR HH 30 APPEARS GREEN IN THIS FALSE-COLOUR IMAGE. A JET OF GAS, IN RED, HAS BEEN VIOLENTLY EJECTED ALONG THE STAR'S AXIS OF ROTATION, ALTHOUGH THE STAR ITSELF REMAINS HIDDEN WITHIN THE DARKEST PART OF THE DISK.

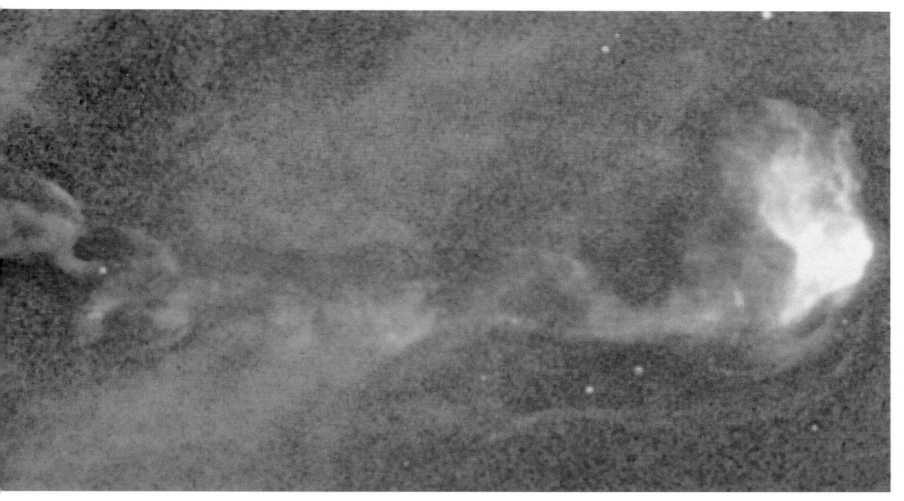

1994. The star itself is still hidden inside the blanket of gas, but indicates its presence by two powerful jets of gas that it is ejecting.

Nebulae are thus true stellar nurseries. They may even create hundreds of stars simultaneously, along a front that is some tens of light-years across. In the densest regions of the Milky Way, whole clusters of stars are formed. These are easy to detect in the sky, because these clusters lie exactly in the plane of the galactic disk, which is the most prolific region of the Galaxy. They all consist of young, bright stars, and their chemical composition indicates that they all have essentially the same age. About 1000 of these clusters are known in our Galaxy. Each contains between ten thousand and several hundred thousand stars. They do, of course, eventually vanish, after a few tens of millions of years, when their stars disperse, scattered in various directions. The Sun itself was probably born along with a few dozen other stars. Since its birth some five billion years ago, however, our star, carried along by the currents of the Galaxy, has covered some four million light-years. We shall never know where the Sun's siblings have gone.

THE PEACEFUL LIFE OF DWARF STARS

When a star is born, the course of its future existence depends solely on its mass. One might imagine, with good reason, that the largest stars would live the longest, and that the smallest would soon die out. In fact, exactly the opposite occurs. As the mass increases, the gravitational field of a star increases, and with it the

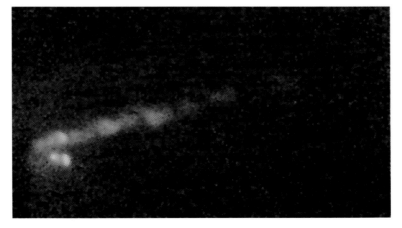

■ HH 34 lies in the constellation of Orion at a distance of 1500 light-years. The infant star regularly ejects puffs of gas at speeds of some 800 000 km/h. In a few hundred thousand years the process ceases, and the jet will disperse into space.

pressure and temperature of the gas. At the centre of the Sun the temperature reaches 15 000 000°C, but it is twice that value in a star ten times the mass of the Sun, and one hundred times at the core of a red supergiant. Under these conditions, there are more thermonuclear reactions and they are more efficient, so the nuclear fuel is more rapidly exhausted.

The smallest stars known, the red dwarfs, have masses that are about one twelfth of a solar mass. Astronomers have had a lot of trouble in locating these miniature stars. Their brightness is, in fact, extremely low. The smallest stars known, LHS 2924, Gliese 623 B, Wolf 359, and Ross 614 have an intrinsic brightness that is 1/10 000 or 1/100 000 times that of the Sun! To put it another way, if one of these red dwarfs were to replace the Sun, its brightness, here on Earth, would be similar to that of the Full Moon. These stars are called red dwarfs because of their deep red colour, caused by their relatively low surface temperature, which is between 2000 and 3500°C.

The faintest star known, ESO 207–61, was only discovered in 1991, 65 light-years away in the constellation of Puppis, part of the older large constellation of Argo Navis. This extremely faint red dwarf has a brightness that is about one millionth that of the Sun, and astronomers wonder if it is actually a star.

In fact, stars with masses of less than 8% of the Sun's may no longer be considered stars, because their energy is not derived from nuclear reactions. The temperature

■ THE SUPERGIANT STAR NGC 2264 IRS LIES IN THE CONSTELLATION OF MONOCEROS. TWO THOUSAND TIMES THE BRIGHTNESS OF THE SUN, THIS STAR IS EMITTING A VIOLENT STELLAR WIND INTO THE SURROUNDING SPACE, WHERE IT HAS CAUSED THE BIRTH OF FIVE SMALLER STARS SIMILAR TO THE SUN. THIS IMAGE WAS OBTAINED IN JUNE 1997 BY THE NICMOS INFRARED CAMERA ON THE HUBBLE SPACE TELESCOPE.

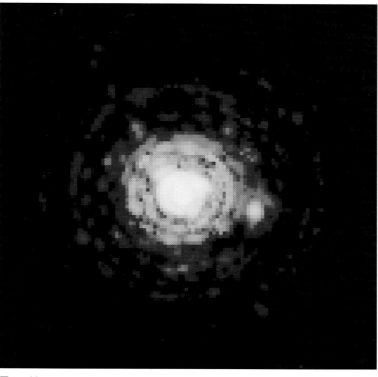

■ HALF OF THE STARS IN THE GALAXY FORM AS BINARIES. GLIESE 623 AND ITS COMPANION GLIESE 623 B ARE JUST 300 MILLION KILOMETRES APART. THIS PAIR OF RED DWARFS LIES AT A DISTANCE OF 25 LIGHT-YEARS. GLIESE 623 B (RIGHT), IS ONE TENTH THE MASS OF THE SUN.

and pressure at their centres are not high enough to initiate atomic fusion. Neither are they planets, because the latter are tiny, cold, and do not radiate their own light, having been formed from the remnants of the disk of gas around newly born stars. A new class of objects was therefore proposed in the 1980s for these particular bodies: they are now known as brown dwarfs.

These objects were purely theoretical for a long time, because their very low luminosity means that they are almost invisible, even with the most powerful telescopes. Nevertheless several serious brown-dwarf candidates were announced between 1993 and 1996, such as PPl 15, Tiede 1, and Gliese 229 B. The last, for example, was discovered at the end of 1995, 18 light-years away. Like a red dwarf, Gliese 229 B emits a feeble amount of radiation, primarily in the infrared, because the gas is heated by the enormous pressure in the interior. Its mass is somewhere between 0.02 and 0.05 solar masses, so we have here a sphere of hot gas that is 20 to 50 times as massive as Jupiter, the largest planet in the Solar System, which is one thousandth the mass of the Sun. The surface temperature of this body, which is about the same diameter as Jupiter but is as red as a glowing ember, is no more than 900°C.

For a long time, based on the fact that the greater the mass, the fewer the stars, astronomers thought that brown dwarfs and red dwarfs made up the bulk of the population of the Galaxy. The surveys carried out in 1994 with the Hubble Space Telescope and the 10-m Keck Telescope on Hawaii have enabled the statistics to be refined. Estimates of the number of mini-stars have now been lowered. It is certainly true that there are probably hundreds of billions of these tiny cool bodies in the Galaxy, but their contribution to the overall mass now seems negligible. It does also seem that for some still-unknown reason, practically no stars are formed below about 0.2 solar masses. Altogether, red dwarfs and brown dwarfs probably represent only 15% of the total mass of the Galaxy.

The most numerous stars in the Galaxy are about one fifth the mass of the Sun, and 0.01 and 0.001 times its brightness. Our Galaxy contains several tens of billions of these stars, which are burning their hydrogen at a very slow rate. As a result, they have extraordinarily long lifetimes, some exceeding hundreds of billions of years, more than twenty times the current age of the Sun. In fact, astronomers are nowhere near observing one of these small red stars towards the end of its life. They can only assume that once all the hydrogen in their cores has been consumed, these stars will shrink and slowly cool until they become invisible.

THE FATE OF RED GIANT STARS

The life of more massive stars, like the Sun and the billions of similar stars, is more eventful but also much shorter. Stars that are slightly smaller than the Sun (about 0.8 solar masses) burn hydrogen in their centres for some twenty billion years. Stars like the Sun last about ten billion years, while stars twice as massive as the Sun have a life expectancy of less than one billion years. In the Galaxy, 90% of the stars are currently busy transforming

■ IN 1995, THE HUBBLE TELESCOPE PHOTOGRAPHED A BROWN DWARF, GLIESE 229 B. THE BODY, ONE TWENTIETH TO ONE FIFTIETH THE MASS OF THE SUN, IS APPROXIMATELY IN THE CENTRE OF THE IMAGE. TO THE LEFT IS GLIESE 229, TOGETHER WITH A DIFFRACTION SPIKE, CAUSED BY THE SPACE TELESCOPE'S OPTICAL SYSTEM.

hydrogen into helium. Astronomers say that these stars lie on the Main Sequence of the Hertzsprung-Russell Diagram, which shows the relationship between the surface temperature of stars and their intrinsic luminosity (see Appendix 3).

Inside the core of a star like the Sun, however, this thermonuclear fusion is extremely efficient, and eventually there is insufficient hydrogen to maintain the same rate of fusion. In the star's core, where hydrogen once predominated – there is 70% hydrogen and 28%

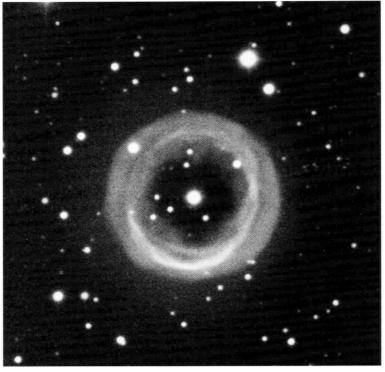

■ THE PRETTY PLANETARY NEBULA SHAPLEY 1 WAS SHED BY A RED GIANT STAR A FEW THOUSAND YEARS AGO. ONCE THE VARIOUS OUTER LAYERS OF THE STAR HAD BEEN EJECTED, ALL THAT REMAINED WAS A TINY WHITE DWARF, WHICH MAY BE SEEN HERE IN THE CENTRE OF THE EXPANDING NEBULA.

helium in a young solar-type star – the balance slowly tips in favour of helium. After five billion years the composition of the centre of the star has altered: there is now just 36% hydrogen, as against 62% helium. Hydrogen-burning can continue for about another four billion years, but it will slowly fade away for lack of hydrogen. But the star's core, which up to now was in perfect equilibrium, counteracting the extreme gravitational force with an equal radiation pressure, is no longer able to

support its own weight, because of the lessened gas pressure. The star's core slowly collapses, and the hydrogen-burning reactions migrate into a thin overlying shell. As the inert core becomes hotter and denser and contracts, the star's atmosphere becomes more tenuous and expands. The star becomes a red giant, an enormous body, 100 times the size and 1000 times the luminosity of the Sun. When the Sun reaches this red-giant stage, in some 4 to 5 billion years, its outer atmosphere, measuring more than 100 million kilometres across, will have swallowed up the planets Mercury and Venus, blown away the Earth's

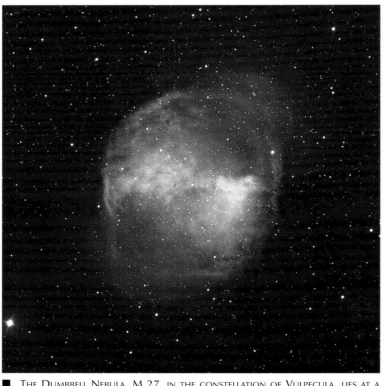

■ THE DUMBBELL NEBULA, M 27, IN THE CONSTELLATION OF VULPECULA, LIES AT A DISTANCE OF 1000 LIGHT-YEARS. ITS EXTERNAL ENVELOPE, WHICH IS FOUR LIGHT-YEARS ACROSS, IS EXPANDING AT A RATE OF 20 KM/S. THE WHITE DWARF AT THE CENTRE IS THE SIZE OF THE EARTH, AND ITS SURFACE TEMPERATURE IS SEVERAL TENS OF THOUSANDS OF DEGREES.

atmosphere, and charred its surface.

In our immediate surroundings, several stars once similar to the Sun have reached the red-giant stage. All appear very bright to the naked eye and the reddish hue means that they stand out from the others, such as Arcturus in Boötes, 37 light-years away, or Aldebaran in Taurus, at a distance of 65 light-years.

THE THROES OF PULSATING STARS

In the still-contracting core of the red giant the temperature has now risen to 100 000 000°C, a critical stage that allows a new thermonuclear reaction to come into play. Several helium atoms fuse to create nuclei of carbon and oxygen. The star is hard put to adapt to all these rapid changes. The red-giant stage lasts less than 10% of the time that the star has spent burning hydrogen. In the case of a star with the same mass as the Sun, the red-giant phase lasts about one billion years. At the centre of the star the nuclear reactions decrease, because the oxygen and carbon produced by the helium fusion are not able to fuse themselves, as the temperature is too low. Surrounding this almost inert core there are helium-burning and hydrogen-burning shells. Under these conditions, instabilities set in, and the star begins to oscillate, with cycles of expansion and contraction. Not only does the diameter alter, but the temperature of the star also varies.

This phenomenon of pulsation may produce some spectacular changes in brightness. There are various classes of pulsating stars, long-period variables, Cepheids, W Virginis stars, and RR Lyrae stars to name

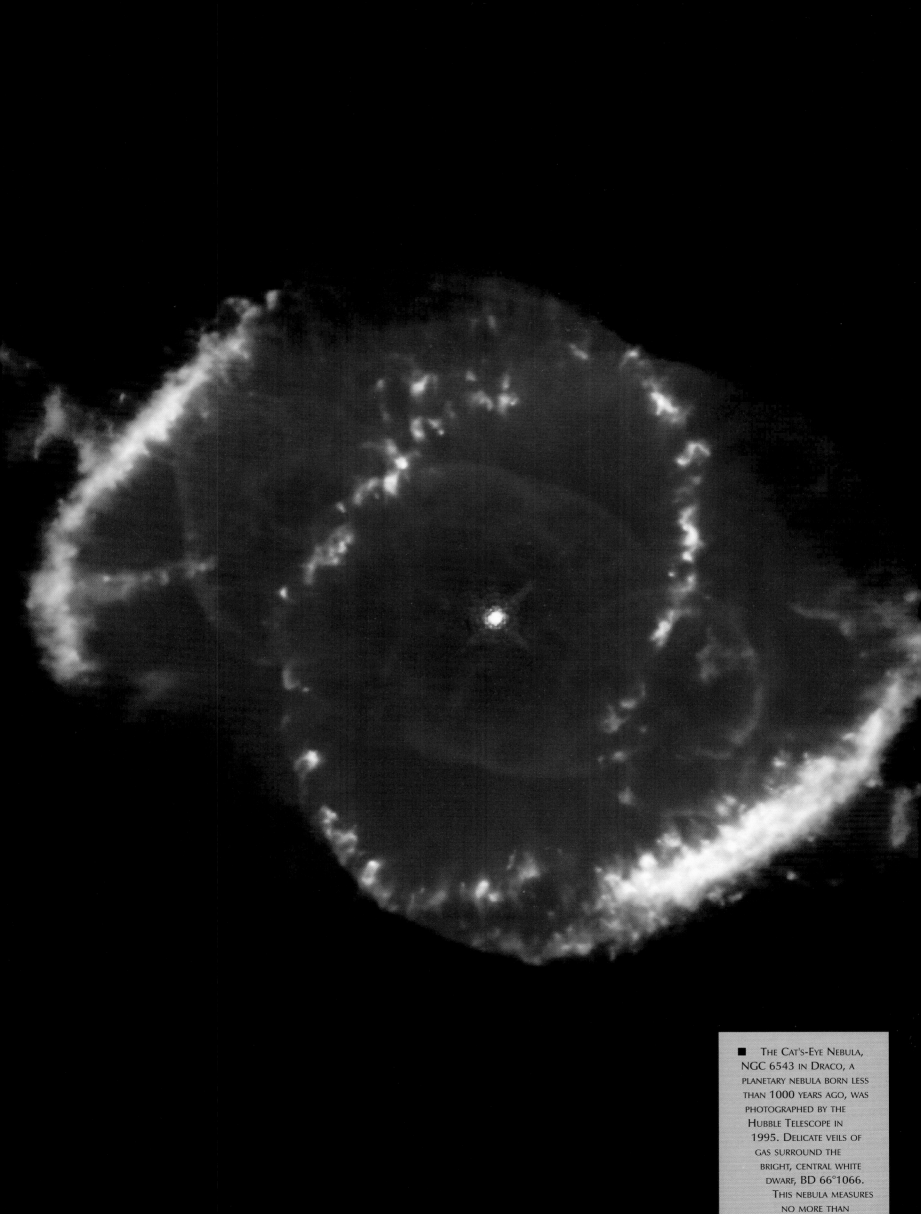

THE CAT'S-EYE NEBULA, NGC 6543 IN DRACO, A PLANETARY NEBULA BORN LESS THAN 1000 YEARS AGO, WAS PHOTOGRAPHED BY THE HUBBLE TELESCOPE IN 1995. DELICATE VEILS OF GAS SURROUND THE BRIGHT, CENTRAL WHITE DWARF, BD 66°1066. THIS NEBULA MEASURES NO MORE THAN THREE LIGHT-YEARS ACROSS.

■ The Helix Nebula was photographed in 1996 by the Hubble Space Telescope. At this large scale, only a small part of the nebula, approximately two light-years across, is seen. Hundreds of gaseous filaments, ejected some 10 000 years ago, can be seen in this extraordinarily clear image. Each of these comet-shaped objects is more than 100 billion kilometres long.

but a few. The last three classes are yellow giants and supergiants, rather than red giants. Their brightness may change by a factor of 2, 10, 100 or even more, with periods that extend, depending on their type, from a few hours to several tens of days.

The most famous pulsating red giant is probably Mira Ceti, which gave its name to the Mira stars, long-period variables, more than 5000 of which are now known. This star has a somewhat irregular period of about eleven months, and changes its luminosity by a factor of 10 000. It lies 400 light-years away, readily seen with the naked eye when at maximum brightness, but requiring a small telescope to be seen at its faintest. Mira measures more than 800 million kilometres across, i.e. about 500 times the diameter of the Sun. During each of its pulsations, the diameter of its gaseous atmosphere varies by around 100 million kilometres. Astronomers believe that Mira-type stars represent a late stage in the evolution of a red giant.

Inside the low-density atmosphere of a Mira star – the gas is about one hundred-millionth the density of water – the material is subject to periodic motions that increase in amplitude such that the material begins to escape. It is driven off by radiation pressure and also by actual waves that are created in the gas by the star's heartbeat. The longer the process goes on, the more violent and irregular the pulsations become, and eventually thousands of billion billion tonnes of matter are shed at a speed of around 10 km/s. The stellar wind becomes a raging stellar gale and, after several centuries or millenia, a beautiful shell of glowing gas begins to appear around the star, like an expanding bubble carrying off a large portion of the star's mass. These envelopes that red giants eject, and which were mistakenly called planetary nebulae when they were discovered in the 18th century, measure two to four light-years across. They disperse into interstellar space after a few tens of thousands of years.

The red giant itself has disappeared. It has shed between a quarter and half its mass in the envelope, and only its tiny, dense, and very hot core remains, in the centre of the nebula. This is the source of the radiation that causes the nebula to glow.

FROM RED GIANTS TO WHITE DWARFS

The dying star has passed from the red-giant stage to that of a white dwarf. This is about the size of the Earth, but its surface is extremely hot at a temperature of between 10 000 and 100 000°C. White dwarfs – the naked cores of stars – have masses that may amount to a few tenths of a solar mass, but which may reach as much as 1.4 solar masses. The matter that makes up a white dwarf is called degenerate. It is solid, incapable of supporting any nuclear reactions, and resembles a gigantic inert crystal. Its density is enormous. If it were possible to remove one cubic centimetre from the interior of a white dwarf and bring it back to Earth, this thimbleful of material would (literally) weigh a tonne!

There are probably 10 billion white dwarfs in the Galaxy, and every year as a red giant becomes extinct, the catalogue of these strange stars grows longer. Although hot and bright when they first appear where there was once a red giant, they are nevertheless extremely hard to detect through a. telescope, because they are not only tiny, but they eventually emit very little radiation. As it loses its gaseous atmosphere and becomes a white dwarf, the red giant shrinks in size from a diameter of 100 million kilometres to just 10 000 kilometres. Astronomers have already discovered several thousand white dwarfs in the Milky Way, most of them recently, and particularly easily when they are still exciting the beautiful colours of planetary nebulae. Over the course of time, white dwarfs cool and fade, until they eventually become lost in the darkness of space.

■ THE HELIX NEBULA, IN THE CONSTELLATION OF AQUARIUS, IS THE CLOSEST PLANETARY NEBULA TO US, AT A DISTANCE OF 450 LIGHT-YEARS. ITS CENTRAL STAR, A VERY HOT WHITE DWARF, IONIZES AN INNER SHELL OF OXYGEN, WHICH APPEARS GREEN, AND AN OUTER, PINKISH SHELL OF HYDROGEN.

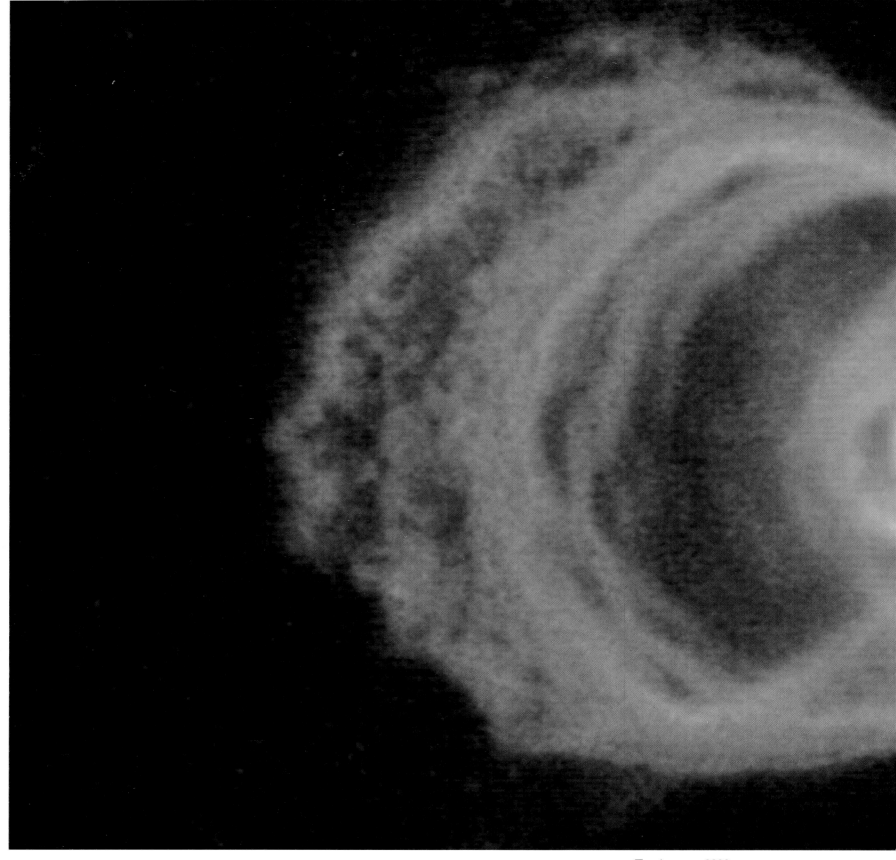

Although stars may come to the end of their lives in the slow extinction of white dwarfs, their flamboyant and dramatic evolution into red giants ensures that they will have descendants, or at least they will bequeath to the Galaxy the promise of a new harvest of stars. During their short period of instability, these solar-type stars shed a large part of their mass into space, not just in the form of hydrogen, but also as heavier elements such as helium, carbon, and oxygen. The atoms ejected in the blue and red expanding stellar shells return to interstellar space, where they will seed other nebulae. The latter, enriched with these atoms from dead stars, will incubate a new generation of stars, which will eventually be born, either as a result of the passage of a galactic spiral density wave, or through compression by the shockwave from the catastrophic explosion of a supernova. ■

■ AT ABOUT 8000 LIGHT-YEARS DISTANCE, THE NEBULA MYCN 18, HAS A DOUBLE SHELL OF RAPIDLY EXPANDING GAS. ON THIS IMAGE OBTAINED BY THE

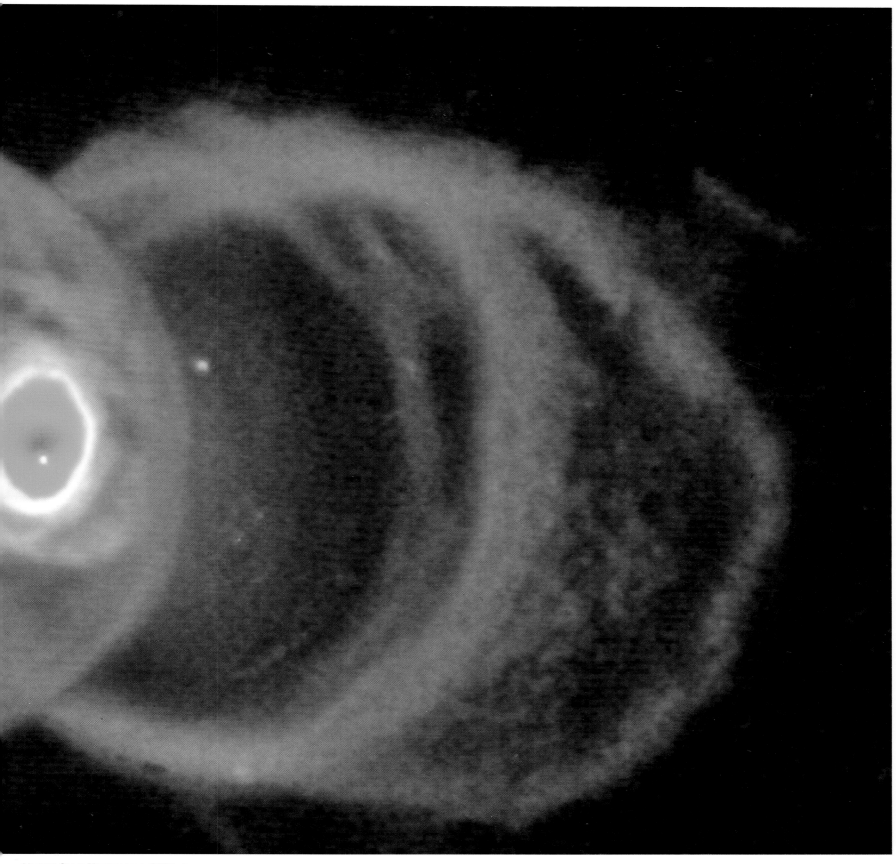

HUBBLE SPACE TELESCOPE IN 1996, THE WHITE DWARF
APPEARS STRANGELY DISPLACED, RELATIVE TO THE
APPARENT DYNAMICAL CENTRE OF THE NEBULA.

The next supernova

■ On 24 February 1987, a supergiant star exploded in the Large Magellanic Cloud, a satellite galaxy of the Milky Way. Although at a distance of 170 000 light-years, this supernova was so bright that it could be seen with the naked eye for several weeks. At its maximum brightness, the supernova was several hundred million times brighter than the Sun. Such an extremely rare event has not been observed in our own Galaxy since 1604.

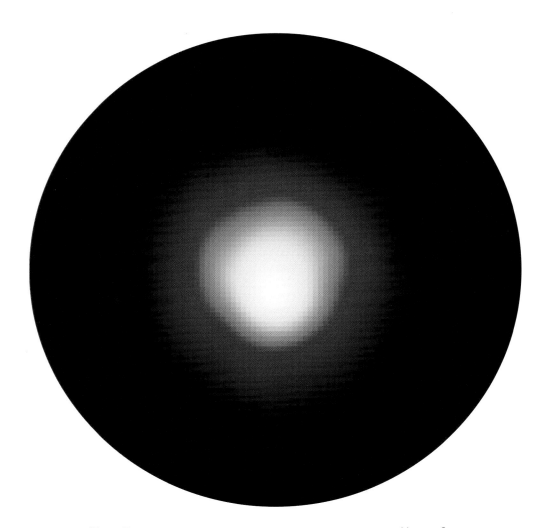

■ BETELGEUSE, WHOSE ACTUAL SURFACE IS SEEN HERE BY THE HUBBLE SPACE TELESCOPE. THIS RED SUPERGIANT IS MORE THAN 500 MILLION KILOMETRES IN DIAMETER, WHICH WOULD SWALLOW UP THE INNER SOLAR SYSTEM AS FAR AS THE ORBIT OF JUPITER.

When night falls in the middle of northern winter, the constellation of Orion culminates above the southern horizon, surrounded by the constellations of Canis Major, Canis Minor, Taurus, Auriga and Gemini. This region is illuminated by eight of the brightest stars in the sky, genuine jewels among the scattered stars of the Orion Arm, the pale band of stars that represents the region of the disk of the Galaxy that lies farther out from the Sun. Here the Milky Way is much fainter than in the region of Scorpius, Sagittarius, or Cygnus, which dominate the summer skies. Orion is, nevertheless, probably the most beautiful constellation in the sky. In northern countries, its appearance in the early light of dawn, in September, reminds us that winter will soon be upon us. This large quadrilateral of bright stars represents the body of the mythological giant who pursued the Pleiades. It is girdled by the three stars of the Belt, Alnitak, Alnilam, and Mintaka, a remarkable close alignment of stars, all bluish in colour and of similar magnitudes. The shoulders of the hunter are marked by Betelgeuse and Bellatrix, whilst his thighs are indicated by Rigel and Saiph. Seen with the naked eye on a cold dark night, Betelgeuse appears ruby red, while the erratic twinkling of Rigel resembles a sparkling sapphire. Although these two stars have such dissimilar colours, they do have one important characteristic in common: they are both exceptionally luminous supergiants. Their remarkable brightness, which makes them appear so close, is actually deceptive. Betelgeuse is, in fact, more than 400 light-years away, and Rigel more than 800 light-years. The latter is one of the most distant stars visible with the naked eye.

Although the Galaxy contains several hundred billion stars, there are less than a thousand that are able to rival in luminosity the two brightest stars in Orion. Deneb in Cygnus, Antares in Scorpius, Eta Carinae, and even Wray 977 owe their extraordinary luminosity to their enormous initial mass, 20, 50, or even 100 times that of the Sun. Gigantic stars, destined to lead brief, flamboyant, and dramatic lives. The rarity of supergiants in the Galaxy is explained by their astonishingly low life expectancy, one-two-thousandth to one-ten-thousandth that of the Sun, i.e. between 1 and 5 million years at the most.

The brightest stars thus live out a mayfly existence. Yet when they are born, supergiant stars follow the same path as stars

■ THE CONSTELLATION OF
ORION INCLUDES SOME OF THE
BRIGHTEST STARS IN THE GALAXY.
BETELGEUSE IS AT UPPER LEFT,
AND THE BLUE SUPERGIANT RIGEL
IS AT BOTTOM RIGHT. THE TRUE
DISTANCE OF THESE TWO STELLAR
LEVIATHANS WAS DETERMINED BY
THE EUROPEAN HIPPARCOS
SATELLITE IN 1997:
BETELGEUSE LIES AT A
DISTANCE OF 400 LIGHT-
YEARS AND RIGEL AT
800 LIGHT-YEARS.

like the Sun. For 90% of them their short existence proceeds normally, with fusion of their central hydrogen, and then of their helium. A star with a mass less than eight times that of the Sun is unable to carry out the nucleosynthesis of heavy elements – those beyond carbon and oxygen. After this stage, it becomes a pulsating star and its brightness varies. It then ejects some of its mass into space, before fading gradually as a white dwarf.

In the most massive stars, however, the core is subject to much higher gravitational force. When they reach the red supergiant stage, it continues shrinking, leading to an increase in the temperature and pressure of the gas. Other nuclear reactions are initiated in deeper and deeper layers of the star. From the outside all that can be seen of such a star is an outer envelope that is about half the temperature of the Sun's surface, but whose size is enormous. Betelgeuse measures about one billion kilometres in diameter, and its true brightness is 10 000 times that of the Sun.

Beneath this cool outer atmosphere, the first layer of gas, at about 3 000 000 to 10 000 000°C, burns hydrogen and turns it into helium. Below that, helium, at over 500 000 000°C, is converted into oxygen, carbon, and nitrogen. Closer to the core, where the nuclear furnace is at over one billion degrees, sodium, neon, magnesium, sulphur, calcium, silver, nickel, and silicon. Finally, at the very centre of the star, atoms of iron are created from silicon. Once it has reached this stage, the star has just a few days to live.

In fact, an atom of iron cannot undergo fusion, and when the core of the star is sufficiently rich in iron it ceases to produce the energy necessary to counteract the force of gravity. For one last time, the core of the star collapses. The state of the gas in the core defies any attempt at description. The density of the material, which is at more than 10 billion degrees, exceeds 10^9, i.e. more than 1000 tonnes per cubic centimetre! Within the atoms, the last of the natural barriers formed by the forces of repulsion fall one by one. Electrons are crushed into the nuclei, where they combine with protons, which are spontaneously converted into neutrons. The gas within the star's core becomes like a fluid of particles, all crushed together. Nowhere in the universe is there a denser or hotter region than the core of a supergiant star that is imploding. The density at the centre reaches 10^{15}, i.e. one billion tonnes per cubic centimetre, and the temperature reaches 150 billion degrees.

■ IN THE HEART OF THE MILKY WAY, THE BRILLIANT STARS OF THE CONSTELLATION OF SCORPIUS ARE, LIKE ANTARES, ALMOST ALL SUPERGIANTS. EACH OF THESE STARS, WHOSE REAL LUMINOSITY IS BETWEEN 1000 AND 10 000 TIMES THAT OF THE SUN, WILL BECOME A SUPERNOVA IN A RELATIVELY SHORT TIME.

FROM SUPERGIANTS TO SUPERNOVAE

This final collapse lasts just a few seconds. It comes to an abrupt halt when the core of the star is just a solid block of neutrons. During this brief period of time, however, support for the overlying layers of the star is removed by the vacuum created by the core collapse, and they crash inwards at 30 000 km/s. The enormous mass of the star smashes into the actual core and rebounds in a cataclysmic explosion. The star turns into a supernova. In the explosion, practically all of the star is flung out into space at a velocity that can exceed one tenth the speed of light.

All high-mass stars finish in this way: as a blinding flash of light in a supernova. The energy produced in a few seconds by a supernova is terrifying. It amounts to several hundred times the energy emitted by the Sun over its whole lifetime of some 10 billion years. Supernova explosions ought to occur in the Galaxy at the rate of one every 25 years. Very few have been observed from Earth, however. In fact, supergiant stars lie precisely in the plane of the galactic disk, which is rich in gas and dust, and where the regular passage of the spiral density wave creates new generations of stars. Most supernova explosions are therefore masked by interstellar clouds, which are a very effective filter. Nevertheless, throughout human history at least a dozen supernovae have undoubtedly been seen.

■ Antares is the brightest star in the constellation of Scorpius. This red supergiant brilliantly illuminates space around it, lighting up the veils of dust in its neighbourhood. Antares, which lies at a distance of 700 light-years, is 10 000 times as bright as the Sun. The life expectancy of such a star does not exceed a few million years.

■ THE SUPERGIANT ETA CARINAE IS KEPT UNDER PERMANENT OBSERVATION BY ASTRONOMERS, WHO EXPECT TO SEE IT EXPLODE AS A SUPERNOVA AT ANY TIME. THIS ENORMOUS STAR, 100 TIMES THE MASS, AND ABOUT ONE MILLION TIMES THE LUMINOSITY OF THE SUN, FREQUENTLY HAS SUDDEN INCREASES IN ACTIVITY, AND HAS BEGUN TO EXPEL A PORTION OF ITS GASEOUS ENVELOPE INTO SPACE.

The first descriptions of 'new stars' being seen in the sky date back to 185, 386 and 393 AD, although it is possible that some of these events, which were scrupulously recorded in Chinese chronicles, refer to comets and not to supernovae. The first unambiguous observation of a supernova is found in Chinese, Arab, and European records, which all describe the appearance in the year 1006 of an extraordinarily bright star on the borders of the constellations of Lupus and Scorpius. This supernova, as bright as the Quarter Moon, must have reached an apparent magnitude (m) of –10. No other star of such dazzling brightness has ever been seen from Earth. Less than fifty years later, another supernova appeared in the constellation of Taurus. As bright as Venus seen with the naked eye, the supernova of 1054 reached magnitude –5. Nine centuries later, astronomers continue to observe the effects of this cataclysmic explosion, which occurred 6000 light-years away. In fact, at the exact position of the explosion in Taurus, accurately described by the Chinese chroniclers, lies the famous Crab Nebula, the tattered remnants of the dead star, and which continue to expand at more than 1000 km/s. The next, less spectacular supernova exploded in 1181. Then, nearly four centuries later, the Danish astronomer Tycho Brahe was lucky enough to observe the supernova of 1572 in the constellation of Cassiopeia, which was also as bright as Venus. Finally, in 1604, Johannes Kepler observed the naked-eye supernova in Ophiuchus, which was as bright as the

■ THESE TWO SUPERGIANTS ARE ILLUMINATING THE NEARBY INTERSTELLAR MEDIUM. WHEN THEY EXPLODE AS SUPERNOVAE, THESE STARS WILL RELEASE MOST OF THEIR MASS INTO SPACE. THE INJECTION OF NEW MATERIAL AND THE SHOCKWAVE FROM THE EXPLOSION WILL LEAD TO THE BIRTH OF NEW STARS IN THE NEBULAE AROUND THEM.

planet Jupiter, reaching an apparent magnitude of about –2.5.

Since then... nothing. Curiously, no other supernova has been observed in the Galaxy for nearly four centuries, during the time, in fact, since Galileo's first telescope of 1609, and as ever more powerful optical instruments have been developed. Recently, astronomers have examined Cassiopeia A, a source of very intense gamma-rays, X-rays, and radio emission, which is also expanding rapidly. By accurately measuring its rate of expansion, which is around 10 000 km/s, they determined that it is the remnant of a supernova that exploded around 1660 in the middle of the Milky Way, in Cassiopeia. But this explosion took place unseen; the flash of light from the supernova was probably so greatly absorbed by the interstellar medium that it was not spotted by 17th-century astronomers.

Today, astronomers are still waiting for the next galactic supernova. To pass the time, they study events that occur in other galaxies. Something over one thousand of these stellar explosions have been recorded, and they have been detected more or less throughout the universe. The first extragalactic supernova was detected in 1885 in the Andromeda Galaxy, at a distance of 2.5 million light-years. At the time, the nature and distance of the galaxies, the nuclear energy processes in stars, and (of course) the origin of supernovae, were all quite unknown. Some spiral galaxies, such as M 83, M 100 or NGC 6946, have been particularly prolific,

■ THIS IS ONE OF THE LARGEST STARS IN THE GALAXY. DISCOVERED IN SAGITTARIUS WITH THE HUBBLE SPACE TELESCOPE'S NICMOS CAMERA, THIS OBJECT, WHICH IS 100 TIMES THE MASS OF THE SUN, RADIATES AS MUCH ENERGY IN SIX SECONDS AS OUR OWN STAR DOES IN A YEAR. THIS SUPERGIANT HAS ALREADY SHED A LARGE PART OF ITS MASS INTO SPACE.

producing between four and six supernovae in just a few decades.

THE VIOLENT EXPLOSION OF TYPE I SUPERNOVAE

Astronomers have obtained valuable information from these 1000 supernova explosions. In particular, they have been able to determine that there are two classes of supernovae. Type II supernovae, as we have just seen, mark the end of supergiant stars, whose initial masses were more than eight times that of the Sun. Type I supernovae are far more violent, and arise in massive binary systems. About half of the stars in the Galaxy exist in pairs, that is to say they are bound together gravitationally, orbiting one another – or rather, orbiting their common centre of gravity. Some pairs of stars are extremely close, where the distance between the two components does not exceed a few tens of millions of kilometres – the distance that separates the Sun and Mercury. When such a binary consists of a massive white dwarf and a star that has evolved to the red-giant stage, a stream of material may be drawn from the expanding envelope onto the white dwarf. The latter's mass increases, and the thermonuclear reactions, which ceased long ago because of lack of fuel, suddenly re-ignite. The reaction is immediate and extremely violent: explosive carbon-burning occurs, and the star explodes – often being completely destroyed – flinging material out into space. Astronomers do not know of any events in the universe that rival the explosion

of Type I supernovae for violence. In a few minutes, the exploding star reaches an absolute magnitude of around -19 – as bright as 10 million Suns! – before fading away over several months, leaving behind a rapidly expanding nebular remnant. With the giant telescopes available at the end of the 20th century, supernovae have been detected at distances of several billion light-years. The current record is held by a supernova discovered in 1997 in a galaxy that lies approximately five billion light-years away.

Although they are still waiting for a galactic supernova, astronomers were lucky enough, a decade ago, to observe a supernova explosion in a neighbouring galaxy, the Large Magellanic Cloud. On 24 February 1987, a new star, easily visible to the naked eye, appeared in the Large Magellanic Cloud, which lies at a distance of 170 000 light-years. Every telescope in the Southern Hemisphere observed the explosion, and immediately detected ejection velocities of 30 000 km/s in the spectra of the expanding cloud of gases. For the first time in the history of astronomy, it was possible for scientists to follow the whole course of the explosion directly – the luminosity of the star increased by more than 100 000 times in just a few hours! – and to scan the archives for details of the star before it exploded. It was not difficult to locate the condemned star on plates taken before 24th February. It was Sanduleak –69°202, a blue supergiant of about 20 solar masses,

■ THERE WAS NO INDICATION WHATSOEVER THAT THE SUPERNOVA OF FEBRUARY 1987 WAS ABOUT TO EXPLODE. IN A FEW HOURS, THE STAR SANDULEAK –96°202

56

about 100 000 times as luminous as the Sun, and about 60 million kilometres across. Since then, the 1987 supernova has been kept under observation by astronomers. Over the course of the years, they have been able to watch its decline in luminosity, the effect of its extreme luminosity on the interstellar medium, and the rapid expansion of the shell of gas that it has ejected.

Neutron stars and pulsars

But what remains behind when stars explode as supernovae? In some cases the cataclysm is so violent that the star is completely destroyed, and disperses into space in a few tens of thousands of years.

Other supernovae leave behind the core of the original supergiant star. The more massive cores of these stars have been compressed far beyond the density of white dwarfs (10^6) that mark the end-points of stars like the Sun. The stellar remnants of supernovae are even stranger bodies than white dwarfs: although their masses lie between 1.4 and 3 solar masses, they are only about 10 km across! The corresponding density is almost unimaginable: 10^{15}. Whatever is a cubic centimetre of material like that would weigh one billion tonnes? These hyperdense stars, which almost entirely consist of a neutron superfluid, are known as neutron stars. Such stars have another spectacular property: the speed at which they rotate.

Originally the supergiant would have rotated about once a month, but as it collapsed its speed of rotation increased as its diameter decreased, just as an ice-skater spins faster as she pulls in her arms. Neutron stars rotate at between 0.25 and 660 revolutions a second! At this frantic speed, these cores of dead stars, which have immense magnetic fields, act as giant dynamos. Some emit charged particles from their magnetic poles, giving rise to radio emission. Others, in binary systems, may even emit X-ray or gamma-ray radiation. It is these invisible forms of radiation that have led to the discovery of neutron stars, which are far too small to be seen by conventional optical means.

In 1967, radio astronomers at Cambridge Observatory in the United Kingdom accidentally identified the first pulsar, a powerful source of radio waves, which were emitted at regular intervals of 1.33 seconds. Pulsars, which are bizarre objects, were soon identified as neutron stars. Their radio emissions correspond to what one might expect from a narrow beam of radiation, created by charged particles lying near the magnetic poles, which sweeps round in space like the beam from a lighthouse. In 1968, to confirm this theory, astronomers pointed their aerials at the Crab Nebula, where the ragged filaments are expanding at a speed of 1000 km/s. In the very centre of the nebula, they located a neutron star, spinning 30 times a second, at the site of the supernova that exploded in 1054.

According to the physicists, neutron stars may form only

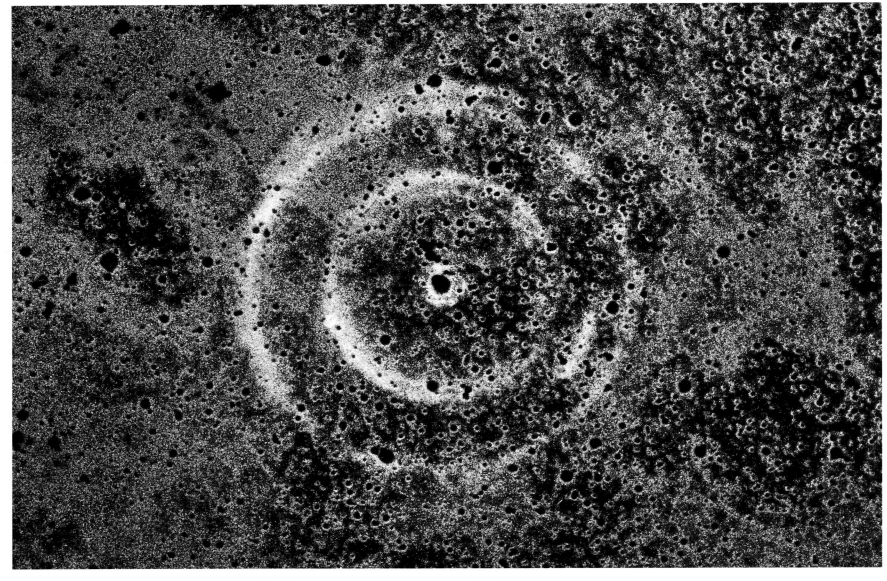

■ Two years after the explosion of the supernova in the Large Magellanic Cloud, the star, Sanduleak −69°202, was practically invisible. The intense pulse of light, however, propagating out into space at the speed of light, continued to reflect from material in the interstellar medium.

from stellar cores with masses between 1.4 and 3 solar masses. Below the lower limit, which was calculated by Chandrasekhar in 1931, a dying star will finish its existence as a slowly cooling white dwarf. But what about above the upper limit? Does some other form of matter exist that is even denser than a neutron star, where the particles are literally squashed together, without the slightest space between them? According to the experts in stellar physics, when a star of 50 to 100 solar masses suddenly explodes as a supernova, its core may exceed the critical mass of 3 solar masses that was determined by Oppenheimer and Volkoff in 1939. There is no force that can prevent the centre of the tiny body, which is just a few kilometres

■ The supernova of 1054 in the constellation of Taurus, was observed by Chinese astronomers, who recorded it in their chronicles. The site of the disrupted supergiant star is today marked by the Crab Nebula, which is expanding rapidly.

across, from suffering gravitational collapse into an ever smaller volume and at an ever greater density. What happens defies human understanding: the material of the dead star collapses – in just a fraction of a second – into an infinitely small space! This cosmic anomaly, which should, according to the equations of general relativity, simultaneously have zero dimensions and an infinite density, is obviously invisible. Its gravitational field is so powerful that nothing, not even light, can escape. It is a black hole.

AT THE HEART OF A DEAD STAR: A BLACK HOLE

What does that mean? According to the theory of general relativity, mass curves the framework of space-time. The greater the curvature, the greater the mass involved. For example, a rocket that has a velocity of 11.2 km/s (40 000 km/h) can easily escape from Earth's gravitational pull. To escape from the surface of the Sun, the rocket would need to exceed 600 km/s, and from the surface of a white dwarf, 5000 km/s. The gravitational field of a neutron star, which literally 'folds' space-time, is enormous: escaping from it would require an incredible amount of energy and a velocity of 200 000 km/s. In the ultimate case, that of a black hole, the escape velocity is 300 000 km/s, which is the speed of light. No information, in the broad sense of the term, can escape from a black hole. The latter is bounded by an 'event horizon', the size of which was calculated by Karl Schwarzschild as just 3 km for a black hole with the same mass as the Sun. This event horizon prevents us from seeing the interior of the black hole, which is known to physicists as a singularity. According to the equations of general relativity, the whole of the dead star is confined there, collapsing indefinitely in a zero-dimensional space.

How can we conceive of any real object that corresponds to such an absurd idea? The physicists have no idea. They would like to avoid the trap presented by a black hole, and somehow move beyond Einstein's equations, avoiding the need to resort to the infinite values that inevitably appear in theories about the singularity that are based on general relativity. They believe that space-time, considered by Einstein to be continuous, may be quantized, i.e. exist in discrete units. Space-time, at extremely minute scales – 100 billion billion times smaller than an atomic nucleus! – may posses a structure, and perhaps consist of ephemeral particles. Being both irregular and in constant flux, this space-time framework would prevent infinities from appearing in the description of a black hole. But this theory, which is already known as quantum gravity, and which Einstein himself worked on without success,

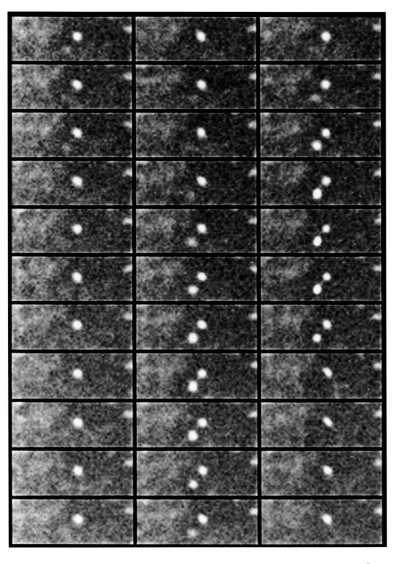

■ THIS FINE SERIES OF IMAGES SHOWS THE VARIATIONS IN BRIGHTNESS OF THE CRAB NEBULA PULSAR, AS OBSERVED BY THE 3.9-M TELESCOPE AT THE KITT PEAK OBSERVATORY IN ARIZONA. THIS STAR FLASHES LIKE SOME INTERSTELLAR BEACON, ROTATING 30 TIMES A SECOND.

has yet to be established. While we await the arrival of this new form of physics, black holes remain objects with properties that are, at best, unknown, and at worst, incomprehensible.

Despite this impasse that has plagued the physicists for decades, astronomers do not, nowadays, doubt the real existence of these voids in space-time; they have begun tracking them down. The first of them was found in 1971 some thousands of light-years away in the constellation of Cygnus, near the star HD 226868. This blue supergiant, which is about 30 times as massive as the Sun, is accompanied by an invisible body at a distance of just 30 million kilometres, with an orbital period of slightly less than a week. This strange companion, known as Cygnus X-1, occasionally emits violent bursts of energy, mainly in the form of X-rays. According to the experts, the giant star's companion is accreting material from its neighbour, and it is this material, accelerated by the gravitational field to relativistic velocities – i.e. to velocities close to the speed of light – that is emitting these violent bursts of X-rays. Only two types of body are capable of producing such energy: pulsars (i.e. neutron stars), and black holes. When they carried out the calculations, astronomers soon discovered that the companion to HD 226868 exceeded the critical mass allowed for a neutron star. It had to be a black hole. In 1980, a second black hole, AO620–00, was discovered in the Galaxy. This object is in the constellation of Monoceros, at a distance of 3200 light-years. There again, astronomers found a binary star. At first sight, this was a perfectly ordinary star, an orange dwarf, slightly less massive than the Sun. At a distance of just 2.2 million kilometres, however, there is a tiny body that is invisible but extremely massive. This black hole is around eight solar masses. Occasionally, violent bursts of X-rays are produced by AO620–00. The black hole distorts the orange star into a teardrop shape, and draws a stream of gas from its outer atmosphere.

Supernovae are of great importance in the evolution of the Galaxy. Their powerful blast expels the equivalent of several solar masses of material out into space to a distance of several tens of light-years. This gas enriches the Milky Way in heavy elements such as oxygen, nitrogen, carbon, and iron. Most of the atoms forming the Earth were created within supergiant stars.

A LARGE-SCALE IMAGE OF PART OF THE VEIL NEBULA (SEE ALSO BELOW) SHOWS SOME OF THE RIBBONS OF GAS THAT REMAIN SOME 100 000 YEARS AFTER A SUPERNOVA EXPLOSION. THE HOT, EXPANDING GAS WILL EVENTUALLY DISPERSE INTO SPACE.

This material initially forms a disk around the black hole's event horizon. From time to time, some of this material falls into this cosmic maelstrom at a velocity of several tens of thousands of kilometres per hour...

About ten invisible black-hole candidates are now known in the Galaxy, from V 404 Cygni to 1E 1740.7–2942, via GS 2000+25. Astronomers wonder what the total number might be. The black holes that have been detected betray their presence by their influence on the stars around them. But how many are dormant, and undetectable, hiding among the spiral arms of the Galaxy? It is possible to make an estimate. In fact, only the largest supergiant stars, with masses between 40 and 100 solar masses, end their lives as a supernova outburst and leave a stellar corpse that is sufficiently massive to collapse into a black hole. Although such stars are extremely rare – there are only a few thousand in the Galaxy today – they evolve extremely quickly, and tens of generations of such stars have existed since the formation of the Galaxy. In addition, although the life-time of a supergiant star does not exceed a few tens of millions of years, that of a black hole is essentially infinite. Scientists therefore estimate that between 10 and 100 million black holes are currently lurking in the disk of the Galaxy.

Where and when will the next supernova explode? The explosion of Sanduleak –69°202 in 1987 shows that these events are un-predictable. When the core of the star starts to collapse, signalling the impending explosion, the star's surface, millions of kilometres above, shows not the slightest sign of the imminent catastrophe. Astronomers know of numerous stars, relatively close, that are likely to explode at any time, which means, on an astronomical timescale, tomorrow, or in 1000 years. Eta Carinae, the largest star in the Galaxy, has been showing signs of instability for more than a century, and is kept under daily scrutiny by astronomers. Another blue supergiant, Wray 977, is beginning to emit a vigorous stellar wind, ejecting its outer gaseous envelope, and uncovering its deeper layers. Along with Eta Carinae and Zeta-1 Scorpii it is one of the most luminous stars in the Galaxy. It is as luminous as one million Suns, and has a mass of around 50 solar masses, and a diameter of some 100 million kilometres. Closer to us, Deneb, Rigel, Antares, and Betelgeuse will one day turn the sky ablaze, appearing brighter than the Full Moon and too bright to look at, and eclipsing all the other stars in the sky for several months, before slowly fading, leaving a gaping hole in the familiar constellation patterns.

Since the origin of our Galaxy, some 13 billion years ago, between 100 million and 1 billion supergiant stars must have exploded, dispersing the rare metallic elements synthesized in their centres into the dark clouds of the nebulae. It is this catastrophic

PART OF THE VEIL NEBULA IN THE CONSTELLATION OF CYGNUS. ASTRONOMERS BELIEVE THAT A SUPERNOVA OCCURS EVERY 25 YEARS IN THE GALAXY. MOST OF THESE GIGANTIC EXPLOSIONS REMAIN INVISIBLE, HIDDEN BY THE DENSE INTERSTELLAR CLOUDS THAT ARE MASSED IN THE GALACTIC PLANE.

stellar nucleosynthesis that explains the 2% of heavy elements found in the heart of the Sun. The latter is far too young, too small, and has too low a mass, to have created any heavy elements itself. Its birth, some 5 billion years ago, was probably prompted, or hastened, by a nearby supernova. The latter left its mark on the Sun, but also, and primarily, on the planets that circle it. Supernovae forged the Earth's iron core, its mantle of silicates, and a considerable fraction of the atmosphere of nitrogen and oxygen that surrounds its surface and protects the species living on it. Without short-lived supergiants and the dramatic supernova explosions, those heavy elements, capable of linking together and forming complex molecules, would not exist. Life would never have appeared on Earth, nor anywhere else in the universe. Even we are children of the stars. ▪

■ TEN YEARS AFTER THE EXPLOSION OF SANDULEAK −69°202, THE REGION AROUND THE SUPERNOVA IS DIFFICULT TO RECOGNIZE IN THIS IMAGE FROM THE HUBBLE SPACE TELESCOPE. THE SPECTACULAR STRUCTURES THAT

SURROUND THE SUPERNOVA ORIGINATE IN MATERIAL
EJECTED BEFORE THE EXPLOSION, WHEN THE SUPERGIANT
STAR EXPELLED A LARGE PART OF ITS OUTER ATMOSPHERE.

Planets by the billion?

■ THE HEART OF THE ORION NEBULA, M 42, PHOTOGRAPHED BY THE HUBBLE SPACE TELESCOPE. THE IMAGE COVERS A FIELD THAT IS ABOUT TWO LIGHT-YEARS ACROSS, AND THE SMALLEST DETAILS VISIBLE ARE ABOUT 6 BILLION KILOMETRES IN SIZE. MORE THAN 700 STARS HAVE FORMED IN THE LAST MILLION YEARS DEEP WITHIN THE ORION NEBULA. OUR OWN SOLAR SYSTEM WAS BORN ABOUT 4.5 BILLION YEARS AGO IN A SIMILAR NEBULA, WHICH HAS NOW DISAPPEARED. THE SUN'S SIBLINGS HAVE BEEN DISPERSED TO THE FOUR CORNERS OF THE GALAXY.

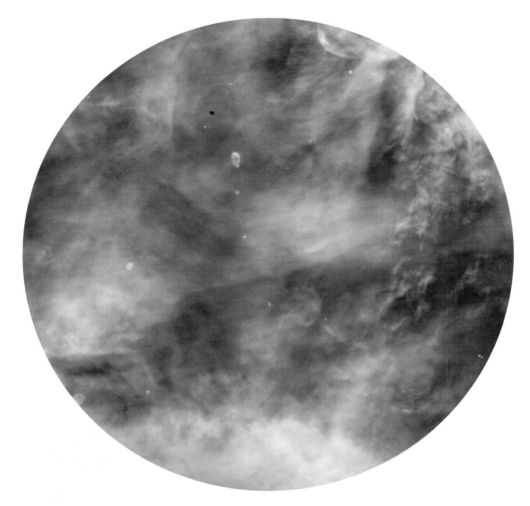

■ In 1994, American astronomers discovered around 150 globules of gas and dust, called proplyds, in the Orion Nebula. A few of them are visible on this Hubble Space Telescope image.

Seen from a distance, as it is slowly carried along by the rotation of the Galaxy, like hundreds of billions of other similar stars, our star appears perfectly undistinguished. Neither very faint nor exceptionally bright, it orbits peacefully and anonymously, some 28 000 light-years out from the galactic centre, between the Orion and Cygnus Arms. On the basis of its physical characteristics, the Sun is an unremarkable star, even though it actually belongs to the 5% of stars that are the most luminous in the Galaxy. The vast majority of stars in the Milky Way are red dwarfs, but the Galaxy still contains more than ten billion stars that are identical to the Sun. Nevertheless, among all this multitude of pinpricks of light that makes up the galactic spiral, the Sun is unique in one respect: it is the only star, in the entire universe, known to possess a planetary system that includes an inhabited planet.

One star and nine planets: that would be the short entry for the Solar System in a hypothetical galactic census. The Sun alone amounts to 99.9% of the mass of the whole Solar System. The largest of the planets, Jupiter, is about 1000th the mass. As for the Earth, it is about one 330 000th the Sun's mass. Starting at the Sun, we find Mercury, Venus, Earth, Mars, Jupiter, Saturn, Uranus, Neptune, and Pluto, all of which, except Pluto, have circular or slightly elliptical orbits lying essentially in the same plane. Mercury lies at just about fifty million kilometres from the star, which floods it with radiation and raises the temperature of its rocky surface to over 430°C. The orbit of Pluto, at about five billion kilometres from the Sun, encloses the Solar System. Here, the influence of the Sun is minimal: the temperature of Pluto's surface does not exceed –220°C. This classic image of the Solar System is, however, both cursory and incomplete: some fifty-odd satellites orbit seven of the planets. Some of these, moreover, could well be considered as planets in themselves. Titan with its thick atmosphere; Io, with intense volcanism; and Ganymede, with complex geology, are all bodies that have masses and sizes that are comparable with those of Mercury or Pluto. Beyond Pluto, the Solar System probably extends for a considerable distance: astronomers suspect that a vast torus exists, populated by billions of comets and small icy planetoids, and which may extend out to nearly one light-year.

■ THIS FANTASTIC NEBULA,
RESEMBLING SOME CREATURE FROM
THE DEPTHS, LIES IN THE
CONSTELLATION OF SERPENS. THE
GAS AND DUST OF WHICH IT
CONSISTS WERE CREATED BY THE
HUNDREDS OF MILLIONS OF
SUPERNOVA EXPLOSIONS THAT
HAVE OCCURRED SINCE OUR
GALAXY WAS FORMED. THE
NEBULA IN SERPENS
RECYCLES MATERIAL FROM
VANISHED STARS INTO
NEW GENERATIONS
OF STARS.

The history of our Solar System

The Solar System has retained traces of its origin. Planets, satellites, and comets were formed 4.5 billion years ago in the disk of gas and dust that gave birth to the Sun. Some of the gas was blown away by the young unstable star, while the silicate particles, which were originally evenly distributed throughout the disk, began to clump together. Like a snowball that grows in size as it rolls down a snow-covered slope, the clumps of particles – planetesimals of a few hundred metres in diameter – slowly grew, gradually sweeping through more and more space, and attracting more and more dust, thanks to their increasing mass. The disk of material surrounding the Sun was thus traversed in all directions by billions of planetesimals that were repeatedly colliding with one another, the larger ones gradually absorbing the smaller. After a few tens of millions of years, the principal planets had formed, with the four largest, known as the giant planets, having accreted the gas left behind by the young Sun, as well as dust. For another billion years, the planets suffered from an incessant bombardment from the hundreds of billions of planetesimals that were wandering throughout the young Solar System. Although the scars have been erased over the course of time on the gaseous planets, and on geologically active bodies, such as Venus, Earth, Io, and Triton, most of the small bodies – Mercury, the Moon, Mars, Callisto, Rhea, etc. – still retain traces, in the form of billions of impact craters of all sizes.

How many stars in the Galaxy have followed the same course as the Sun, and been surrounded by a train of planets? This question has occupied thinkers since the time of Epicurus, who raised it in 300 BC in his *Letter to Herodotus*. Since then, philosophers have argued about it incessantly, some maintaining that the history of the Solar System is unique, others advancing the opposite view that the processes found in the Solar System are

■ Here is another region of the nebula M 16 in Serpens, which lies at a distance of 7000 light-years. This ghostly nebulosity is about six light-months across. Some of the stars that are forming in this region of the Milky Way, rich in interstellar dust, will probably give rise to systems of planets.

universal. Hiding behind this nominally astronomical question, there is a debate with intense philosophical implications: Are we alone in the universe? Has life, on the other hand, developed elsewhere? Above all, are there other intelligent species, other civilizations?

Is the Solar System unique?

One thing is certain: planets are the sole havens for life in the universe; other places, such as stars or nebulae, are either too dense and hot or too empty and cold to allow an elaborate chemistry to develop as far as the complex molecules that form the basis of life. The search for extrasolar planets has therefore naturally always been linked with the search for possible extraterrestrial life.

Until the very end of the second millennium, astronomers have been unable to remove the question of doubt about the existence of other planets. Our Solar System, scrutinized through telescopes and by spaceprobes sent in all sorts of directions, has remained the sole example of a planetary system known in the universe. Tens of billions of stars have been photographed by giant telescopes, but when it comes to planets – none. The reason for this inability to discover extrasolar planets is simple. Stars are enormous spheres of gas that shine in their own right. By contrast, planets are minute, and only feebly reflect light from the stars around which they orbit.

The example of Beta Pictoris

Rather than planets, astronomers found, in 1984, the first signs of the existence of other planetary systems. In that year, the American astronomers Bradford Smith and Richard Terrile discovered a disk of gas and dust around the star Beta Pictoris. Lying at a distance of 62 light-years, in the southern constellation of Pictor, this star is slightly hotter and more massive than the Sun, and is definitely much younger.

Beta Pictoris is not more than 100 million years old. The disk that surrounds it, which we can see almost exactly in profile, measures more than 200 billion kilometres in diameter and is considered by scientists to be a protoplanetary disk identical to the one that gave birth to the Solar System. Numerous complementary indications reinforce this idea. First, the grains of dust, which reach a few tenths of a millimetre in size, could well be mistaken for interplanetary dust particles, whereas dust particles in the interstellar medium are microscopic, like smoke. Second, there is the spectrum of the star itself, which shows the signature of extremely short-lived, violent events, which, to the French astronomer Anne-Marie Lagrange, appear to be the impact of comets with Beta Pictoris. Such impacts, although relatively rare in the Solar System, are doubtless extremely frequent when a planetary system is in the process of formation. Finally, the discovery that the distribution of dust around Beta Pictoris is by no means homogeneous, seems to indicate that planets may have already started to condense and sweep up material from the disk.

The discovery of the disk around Beta Pictoris, which is so close to the Earth that it suggests that such systems ought to be all-pervasive in the Galaxy, gave great impetus to the search for other, identical, disks of dust. After ten years of intensive search, Beta Pictoris is no longer the sole example of a protoplanetary disk. Thanks to the European Infrared Space Observatory (ISO), astronomers have discovered disks around Vega and Fomalhaut, among others, although they remain difficult to observe because, unlike that around Beta Pictoris, they are seen face-on, and not in profile, the latter orientation being particularly favourable for detection.

SEARCHING FOR EXTRASOLAR PLANETS

In parallel with this search for planetary systems that are forming around young stars, astronomers have long hoped to discover their Holy Grail: a true extrasolar planet. Failing a direct observation, which until very recently was considered impossible, they devised indirect methods of detection. The first of these techniques has been tried for some fifty years: it involves measuring the individual paths followed by stars against the sky background as accurately as possible. If stars are isolated in space, their tracks, over a period of a few years, are straight lines a few seconds of arc long. If, on the other hand, one of these stars is accompanied by a planet, the latter's gravitational attraction causes a slight oscillation of the star about its mean position, and affects its path, which becomes sinusoidal. The effect is so weak, however, that so far all attempts to discover extrasolar planets by this method have proved futile.

The second technique consists of closely monitoring the brightness of stars that are suspected of having a retinue of planets, using extremely sensitive photometers. If, during the course of the observations, one of these stars fades slightly and then returns to normal, it is possible that the star was partially eclipsed by a planet. Statistically, such an eclipse is exceptionally rare, and, if it occurs, the drop in luminosity lies between 1/100 00th and 1/100th. Naturally, this method applies only to planetary systems that appear edge-on, as seen from Earth. This means that to have the slightest degree of success, the method needs to be applied to a very large sample of stars, and demands many years, or even decades, of patience from astronomers. Nevertheless, it may already have borne fruit. In 1995, the French astronomer Alain Lecavelier discovered a fade of 3% in the brightness of Beta Pictoris during a photometric programme in 1981. This could be a possible signature of a giant planet, slightly larger than Jupiter, that would be taking one or two decades to orbit the star. To obtain positive confirmation of this extraordinary discovery, it would require continuous surveillance of Beta Pictoris over a period of ten or twenty years...

The third method is based, like the astrometric method, on the gravitational effect exerted by a planet on its star. The slight, cyclic oscillation of the star in its path through space is detected, in this case, by a spectrograph, with is capable of measuring the variations in the star's velocity relative to the Earth. The required precision, which is around 15 m/s, was attained by several teams in both Europe and the United States around the end of the 1970s.

It was this spectrographic method that led to the

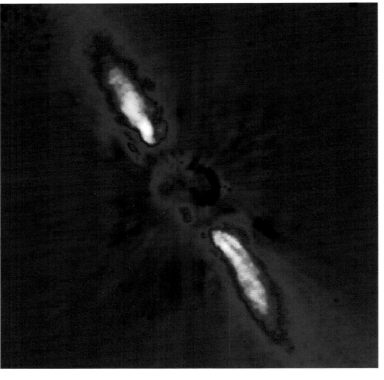

■ THE DISK OF DUST THAT SURROUNDS BETA PICTORIS, PHOTOGRAPHED WITH THE 3.6-M TELESCOPE AT LA SILLA OBSERVATORY IN CHILE. THE STAR ITSELF, OTHERWISE EXTREMELY BRIGHT, HAS BEEN HIDDEN BY A MASK. COMETS, AND PERHAPS ALSO PLANETS, HAVE ALREADY BEEN FORMED IN BETA PICTORIS' DISK.

discovery of the first extrasolar planet. At the Haute-Provence Observatory, Michel Mayor and Didier Queloz, two Swiss astronomers, followed 142 bright stars, that are close to the Earth, and physically similar to the Sun, for a period of several years. In October 1995, they detected a planet of the same mass as Jupiter near 51 Peg, a star that is about 8 billion years old, and which has the same mass and luminosity as the Sun. It lies at a distance of just 45 light-years, in the constellation of Pegasus. The planet, called 51 Peg B, orbits in slightly more than 4 days at a distance of just 7.5 million kilometres. Such close proximity to the star is surprising: Mercury, the closest planet to the Sun, is never less than 46 million kilometres away and is also a completely rocky planet. Astronomers do not know how a giant planet, a large fraction of which is probably gaseous, could form so close to a star. At such a small distance, the planet 51 Peg B must have a temperature of about 1000°C. The theorists have, however, come up with a scenario to explain the strange orbit of 51 Peg B. The planet may have formed, 8 billion years ago, much farther from the star. Subsequently, a process of dynamical braking, caused by the friction between the growing body and the disk of dust surrounding the star, could have caused it to slowly approach the latter.

At the beginning of 1996, events began to happen fast. An American team led by Geoffrey Marcy, using the same technique as Michel Mayor, but at Lick Observatory in California, announced the discovery of two new giant planets, orbiting two nearby stars, 70 Vir and 47 UMa. These two stars are, like 51 Peg, visible to the naked eye, and their physical characteristics are again very similar to those of the Sun. 70 Vir, in the constellation of Virgo, is just 75 light-years away from the Sun. Its planet is about six times the mass of Jupiter and takes about 4 months to complete its elliptical orbit, some 75 million kilometres from the star. As for the star 47 UMa, it lies at a distance of 45 light-years, in the constellation of Ursa Major. 47 UMa B is a planet of three Jupiter masses, lying 315 million kilometres from the star, with a circular orbit that it completes in slightly more than 3 years.

The orbital characteristics

of these planets are very close to those of the planets in our Solar System, because 70 Vir B is about the same distance as Mercury from the Sun, while 47 UMa B is about the same distance in its system as Mars is in our own.

Even more remarkable, the star Upsilon Andromedae, where one extrasolar planet had already been discovered in 1997, has now been shown to be a complete planetary system. By carrying out continuous monitoring of the star, which lies at a distance of 44 light-years, Geoffrey Marcy and Paul Butler have found two new planets! Upsilon And A covers its complete orbit, which is just 9 million km in radius, in only 4.6 days. Upsilon And B orbits at a distance of 125 million km in 241 days, and finally Upsilon And C has an orbit 375 million km in radius, which it completes in about three and a half years.

As we come to the end of the millennium, discoveries are occurring at the rate of one new planet roughly every two months. By mid-1999, astronomers have already compiled a list of some twenty extrasolar planets, most of which are less than 100 light-years away, and generally comparable in mass to Saturn and Jupiter, or rather more massive. In fact, the extremely effective, spectrographic search method introduced by Mayor and Queloz at the Haute-Provence Observatory has only one fault: although it enables one to determine extremely accurately the distance of the planet from its star, its orbital period, and even the eccentricity of its orbit, it is less accurate when it comes to determining the mass. The one factor that observers lack is the orientation of the plane of the orbit of the extrasolar planet relative to us. Without this parameter, estimates of the mass are difficult and uncertain.

In one case, however, researchers do seem to possess full details of an extrasolar planet. The planet 55 Cnc B was discovered in 1997 by Geoffrey Marcy and Paul Butler. It lies about 41 light-years away, in the constellation of Cancer. According to their calculations, the invisible planet may have a mass between 0.8 and 10 Jupiter masses, or even more... Dramatically, the American astronomers David Trilling and Robert Brown discovered that, in addition to the planet, the star 55 Cnc is surrounded, like Beta Pictoris, by a vast disk of gas and dust. (In addition, there

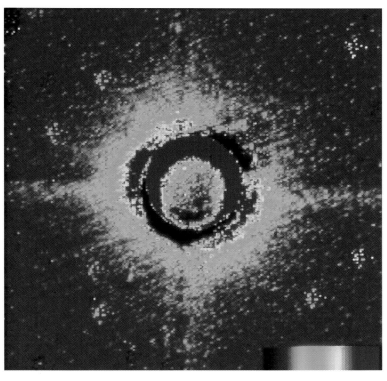

■ THE DISK OF DUST SURROUNDING THE STAR 55 CNC, IMAGED USING A LYOT CORONAGRAPH ON THE NASA 3-M INFRARED TELESCOPE AT THE OBSERVATORY OF HAWAII. THE DISK, COLOURED RED, IS CLEARLY VISIBLE JUST OUTSIDE THE MASK, AT TOP LEFT AND BOTTOM RIGHT.

■ THE LAGOON NEBULA, IN THE CONSTELLATION OF SAGITTARIUS, IS 5000 LIGHT-YEARS AWAY FROM US. IN 1997, ASTRONOMERS AT THE EUROPEAN SOUTHERN OBSERVATORY (ESO), AT LA SILLA, IN CHILE, DISCOVERED A STAR IN THE NEBULA THAT IS SURROUNDED BY A PROTOPLANETARY DISK.

are signs of the presence of ice.) The disk, unlike the planet, is readily visible in the infrared, provided a coronagraph is used to hide the bright star itself. The researchers found that this disk is inclined at about 25° to the plane of the sky. This was just the information that Marcy and Butler lacked to determine the exact mass of the planet orbiting 55 Cnc. All the specialists agree, in fact, that the plane of the orbits of planets agrees with that of the circumstellar disk from which they are born. We now know the mass of 55 Cnc B: twice that of Jupiter. It was a godsend to the specialists because now, for the very first time, they have all the dynamical characteristics of an extrasolar planet: mass (2 Mj), orbital period (14.65 days), semi-major axis (0.11 AU, i.e. about 16 million km), and orbital eccentricity (0.03).

Most of the extrasolar planets discovered as this century draws to a close orbit stars that are very similar in age and luminosity to our own Sun. This certainly does not mean, however, that nature has restricted the formation of planetary systems to just solar-type stars. The proof of this is the spectacular discovery, in 1998, of an extrasolar planet of two jovian masses, orbiting the nearby star Gliese 876. Lying in the constellation of Aquarius at just 15 light-years from the Earth, this star is a red dwarf, one fifth of the Sun's mass, and about one hundredth of its brightness. Red dwarfs are the most numerous stars in the Galaxy. Finding an extrasolar planet around one of them, so close to us, implies that, statistically, the occurrence of planets orbiting red dwarfs is by no means rare. Gliese 876 B orbits its small red star every

■ THE ENVIRONMENT AROUND SUPERGIANT STARS IS NOT SUITABLE FOR THE FORMATION OF PLANETARY SYSTEMS. HERE, ABOUT TEN PROPLYDS, WHICH ARE TOO CLOSE TO THE FOUR STARS OF THE TRAPEZIUM IN ORION, WILL EVAPORATE IN LESS THAN 100 000 YEARS. THEY WILL NOT HAVE TIME TO FORM PLANETS.

60 days, at a distance of about 30 million kilometres.

As we move into the next millennium, it is almost certain that researchers, using giant telescopes, like the Keck telescopes or the VLT, in interferometric mode, will attempt to obtain direct images of the closest or brightest of these extrasolar planets. Succeeding with such a challenge seems somewhat utopian at present, but astronomers' persistence and desire to explore mean that it will be inevitable. All the more so, because they will soon no longer be restricted as to where they should turn their telescopes. One thing is clear from this astonishing series of discoveries, made in just a few years, is that henceforth we can be certain that there are tens of billions of planetary systems in the Galaxy.

In the light of this immense number of potential targets, as the 21st century opens, the search for extrasolar planets will take on a new dimension. Astronomers want to obtain a proper statistical sample, covering tens or hundreds of thousands of stars. In 2002, the small Corot space telescope, built in France by the national space-research centre (CNRS), will be launched. Its mission is to continuously monitor some 50 000 stars for six months at a stretch. Corot will attempt to detect the transit of extrasolar planets across the disks of their parent stars. The depth of the fade and the duration of each 'mini-eclipse' will enable the researchers to calculate the diameter of the planet and the orbital period. The succeeding phase envisaged by ESA scientists is even more ambitious. The GAIA astrometric mission, which may be launched by an Ariane 5 rocket in 2010, will be able to detect practically all the planets with masses equal to, or greater than that of Jupiter, around a million stars within a sphere 1000 light-years across. In the even longer term, both American and European scientists with, respectively, the Terrestrial Planets Finder and the Darwin missions, even expect to discover – if they exist – all the planets resembling the Earth in a sphere 100 light-years across. The Terrestrial Planets Finder and the Darwin missions, which are both space interferometers some fifty metres in diameter, will then study those planets in detail, in the hope of finding liquid water on their surfaces. Finally, and above all else, astronomers dream of detecting in the atmospheres of these highly hypothetical blue planets, an oxygen excess, which might betray the presence of forms of life on their surfaces. In the absence of any external reference, scientists do, in fact, estimate that all the forms of life that may exist in the universe must be organised on more-or-less the same scheme as life on Earth. Carbon will be the atom that forms the basis of molecular bonds and water the universal solvent. As a site for life, the planet will have a stable orbit, and be surrounded by a protective atmosphere, providing a stable climate, and, finally, it will orbit a star that does not vary greatly in brightness.

On this basis, Earth is exceptional in the Solar System. Its distance from the Sun, its mass, and the pressure of its atmosphere have precisely those values that allow water to be present permanently in its liquid phase. Although very close to Earth, the planets Venus and Mars have been transformed, respectively, into a torrid desert, and an icy desert. In 1996, the spectacular announcement by NASA of the discovery of possible fossil life-forms in a meteorite from Mars that had been recovered in the Antarctic, inflamed imaginations. It soon became obvious, however, that the apparent fossil structures, which were 3.5 billion years old, were probably of mineral origin. Most of the other bodies in the Solar System have no atmosphere. Subjected to the vacuum of space, to solar radiation, and to changes in temperature amounting to several

hundred degrees, their surfaces are utterly sterile. Titan, the small 'planet' orbiting Saturn, is protected by an atmosphere of nitrogen similar to our own, and with a slightly higher pressure. But it is −170 to −210°C at the surface. No complex organic chemistry is possible at those temperatures. If we ignore the suggestion – highly unlikely in view of the temperatures and pressures that prevail, and which vary wildly as a result of storms that are large enough to engulf the Earth – of organisms floating in the clouds of the giant planets, there are no other forms of life in the Solar System.

And what about elsewhere? If, as on Earth, any hypothetical extraterrestrial life-form requires hundreds of millions or even billions of years of evolution to develop complex forms, then we need to look for planets that are sheltered by stable stars with long lifetimes. Excluding double stars, variables, and supergiants that explode after just a few million years, there still remains a vast number of solar-type stars, which exhaust their hydrogen in 10 billion years, and red dwarfs, whose ages may, in principle, exceed 100 billion years. In general, however, astronomers and biologists agree in avoiding red dwarfs, because these are undoubtedly too cool and too low in luminosity to heat any planets adequately. Seen from their icy surfaces, which would be plunged in permanent twilight, their stars would, in most cases, be hardly any brighter than the full Moon. In any case, their low masses may prevent the formation of planets around them.

Is it possible that, on some of the several million habitable planets that remain, forms of life may have actually developed? Once again, the only relevant data comes from the Solar System. Just one planet, in a system consisting of some fifty bodies, has developed any forms of life. On the Earth, life arose some 3.8 billion years ago, practically at the very beginning of the planet's existence. At first slowly, and then explosively at the beginning of the Palaeozoic, it conquered all possible and conceivable, niches from the dark oceanic depths, under fantastic pressures, to the high-altitude deserts of the Himalayas and the Andes, where the air is freezing and oxygen is scarce. There is a violent contrast between our planet, teeming with life and

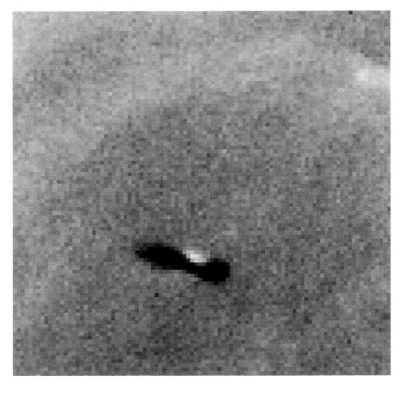

■ THIS PROPLYD IN THE ORION NEBULA, SOME 1500 LIGHT-YEARS AWAY, WHICH RESEMBLES THE DISK AROUND BETA PICTORIS, IS SEEN FROM THE SIDE. THE YOUNG CENTRAL STAR IS PARTLY HIDDEN BY THE GAS AND DUST THAT SURROUNDS IT.

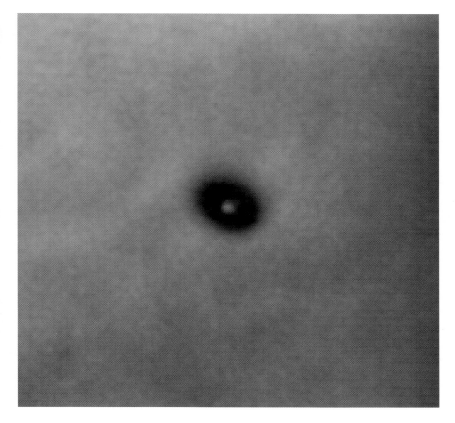
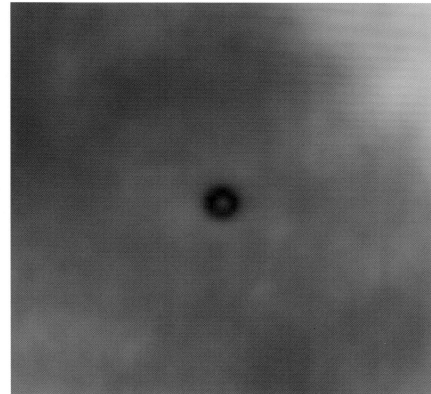
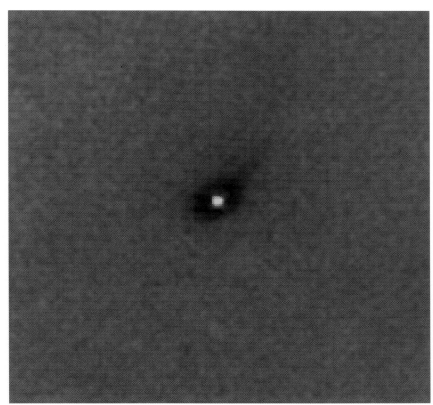
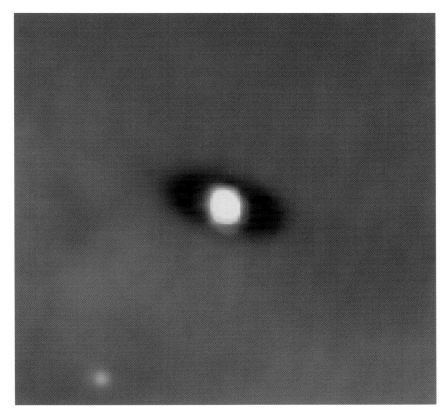

where there are at least a million different species, and the other, lifeless, bodies in the Solar System.

OUR PLANET'S EXCEPTIONAL CIRCUMSTANCES

What was special about our planet from the very beginning, that it should be so endowed with life? Although astronomers had the answer literally hanging over their heads, they did not discover it until the beginning of the 1990s. It was, quite simply, the Moon. The Earth and Moon do, in fact, form the only double planet in the Solar System. The Moon, which is so familiar to our eyes, is not an ordinary satellite like all the others, but is actually a planetary curiosity. Unlike the satellites of Mars, Jupiter, Saturn, Uranus, and Neptune, which are one-five-hundredth to one-fifty-thousandth of the mass of their parent planets, the Moon is just 1/81 the mass of its parent planet. In addition, the Earth and Moon have stabilized one another. This gravitational coupling, experts have discovered, had a crucial effect on the Earth's history. The Moon stabilized the rotational axis of our planet, enabling it to enjoy a constant amount of insolation, so the global climate remained relatively stable. It is probably here that is the secret of success for life on Earth. Without this orbital stability that the Moon has created, life might never have been able to evolve – continuously – over the years, towards more and more complex forms. Already unique for these dynamical reasons, the Earth-Moon pair has another feature that is unique in the Solar System. It lies within the orbital zone in which water can exist in its liquid state. According to some researchers, this zone is extremely narrow,

and extends only from the Earth's orbit to that of Mars. If our planet had been 10 to 20 million kilometres closer to the Sun, its climate would have become like a furnace: 50 million kilometres farther out, and it would be covered in an icy desert. A third factor seems to have helped to ensure the survival of life on Earth. This is the protective presence of Jupiter, the giant planet with a powerful gravitational field. During the more than 3 billion years when life was developing on Earth, Jupiter would have deflected millions of comets and asteroids away from the Earth, either by ejecting them from the Solar System, or by mopping them up. Without the gas giant, our planet would

have been constantly bombarded, with its oceans and continents

devastated by horrendous impacts, and life, repeatedly decimated, would probably never have passed the bacterial stage. This theory was considerably reinforced by Comet Shoemaker-Levy's spectacular series of impacts with Jupiter in July 1994.

So we begin to see the rare and extremely precious nature of our home. We may perhaps owe our existence to an extremely unlikely series of chance events, with an infinitely small overall

EACH OF THESE STARS, SIMILAR TO THE SUN, HAS A
PLANETARY COMPANION.

statistical probability.
Only the astronomical
number of stars in the Galaxy has ensured that it actually came
to pass. We now know that there are other planetary systems
in the universe, but our home may well be unique. The
majority of landscapes of other worlds – even though they may
be reckoned in millions or billions – are probably as empty and
silent as the freezing plains of Enceladus or Callisto. ∎

The enigma at the heart of the Milky Way

■ THE MILKY WAY IS A SPIRAL GALAXY. THE SUN LIES IN THE PLANE OF THE DISK, WHICH CONSISTS OF SEVERAL HUNDRED BILLION STARS, AT A DISTANCE OF SOME 26 000 LIGHT-YEARS FROM THE GALACTIC CENTRE. THIS INFRARED IMAGE, OBTAINED BY THE COBE SATELLITE, SHOWS THE CENTRAL BULGE, WHERE TENS OF BILLIONS OF OLD STARS ARE CONCENTRATED.

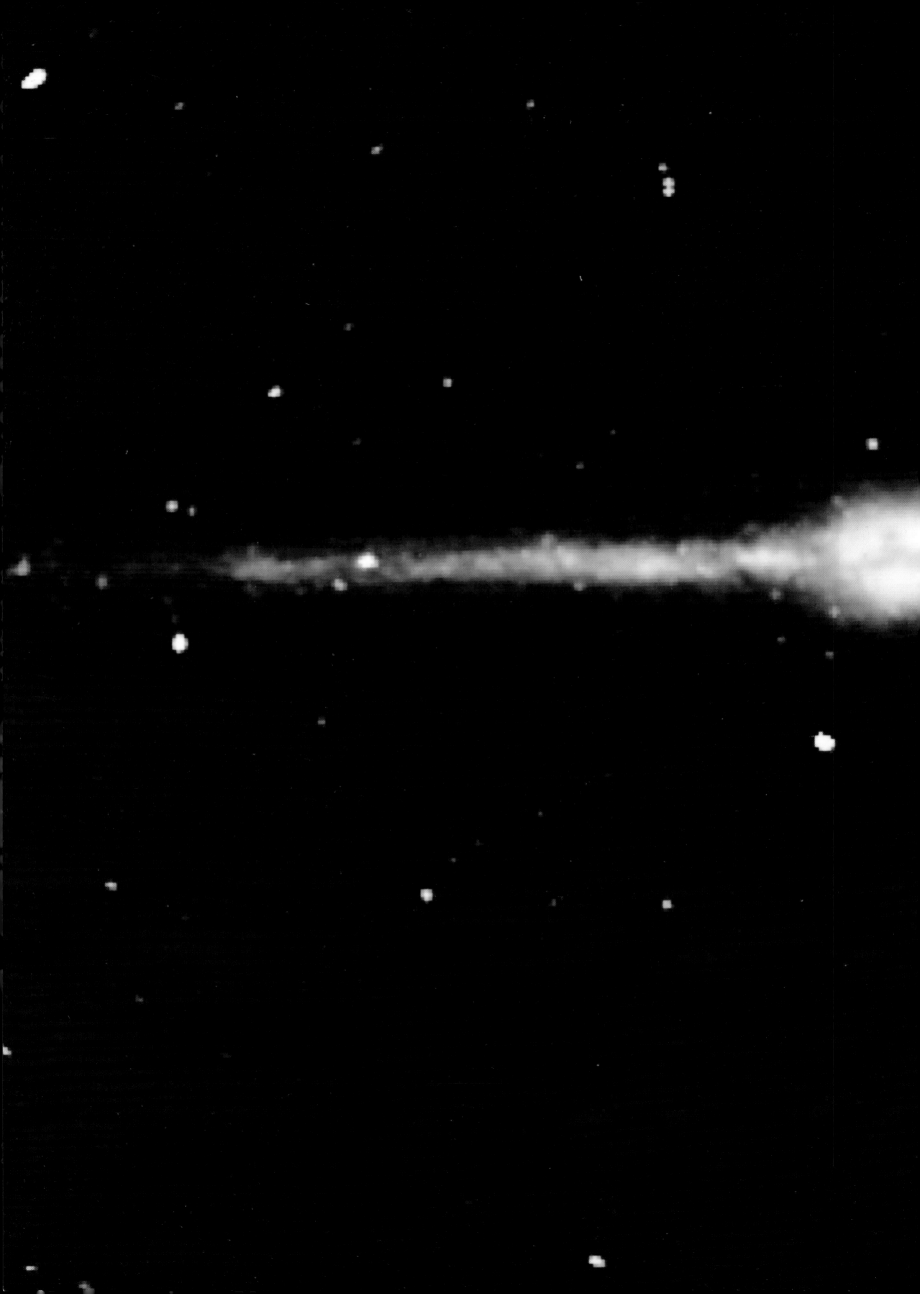

ur Galaxy is revealed in all its splendour when seen from the Southern Hemisphere, from the Australian, Namibian, or Chilean deserts, and beginning at the end of June, at the start of the southern winter. At that time of year, the Milky Way culminates at the zenith, crossing the sky from one side of the horizon to the other. When the nights are clear and the freezing, dry air is absolutely pure, this long sparkling ribbon is bright enough to faintly illuminate the landscape. Every year at around this time, enthusiastic astronomers at the Andean observatories carry out a strange, enchanting experiment. They lie down on the ground and stare up at the sky, trying to take it all in at a single glance. After doing this for several minutes, all of a sudden their perception changes, and they see the Galaxy in its true perspective. Behind the foreground stars that make up the constellations, they see that the stars are not simply the flat ribbon of the Milky Way, but that they stretch away towards the centre, giving a three-dimensional view of our Galaxy as seen from the side.

Their attention is drawn, irresistably, towards the zenith, where the constellations of Ophiuchus, Scorpius, and

Sagittarius wheel slowly across the sky. The beauty of this region of the Milky Way is enough to take your breath away. With just the naked eye you can see the diffuse bulge of the Galaxy, divided into two by the dark lanes of the disk, which is sprinkled with nebulae and clumps of stars.

It is here, on the borders of Ophiuchus and Scorpius, and just within Sagittarius, that the centre of our island universe, the focus of the 1000 billion stars in the Milky Way, lies hidden. It is the most eagerly sought-after object in the entire Galaxy, yet it is also the most mysterious, because it is invisible. If one turns a telescope on this region of the sky, all that can be seen is a dense cloud of stars. On photographs they appear to be touching one another, whereas the distances between them are, in fact, just the same as those that separate the Sun from the nearest stars: several light-years. The line of sight towards the galactic centre, 26 000 light-years long (i.e. 260 billion billion kilometres), lies exactly in the plane of the disk, which is not only clogged with stars but also with clouds of interstellar dust, which absorbs starlight. The central region of the Milky Way is hidden behind an almost perfect filter, which only transmits about one million billionth of the light that the centre

■ The galactic centre lies in the constellation of Sagittarius, in the centre of this image. In this standard photograph taken at optical wavelengths, however, the nucleus is quite invisible.

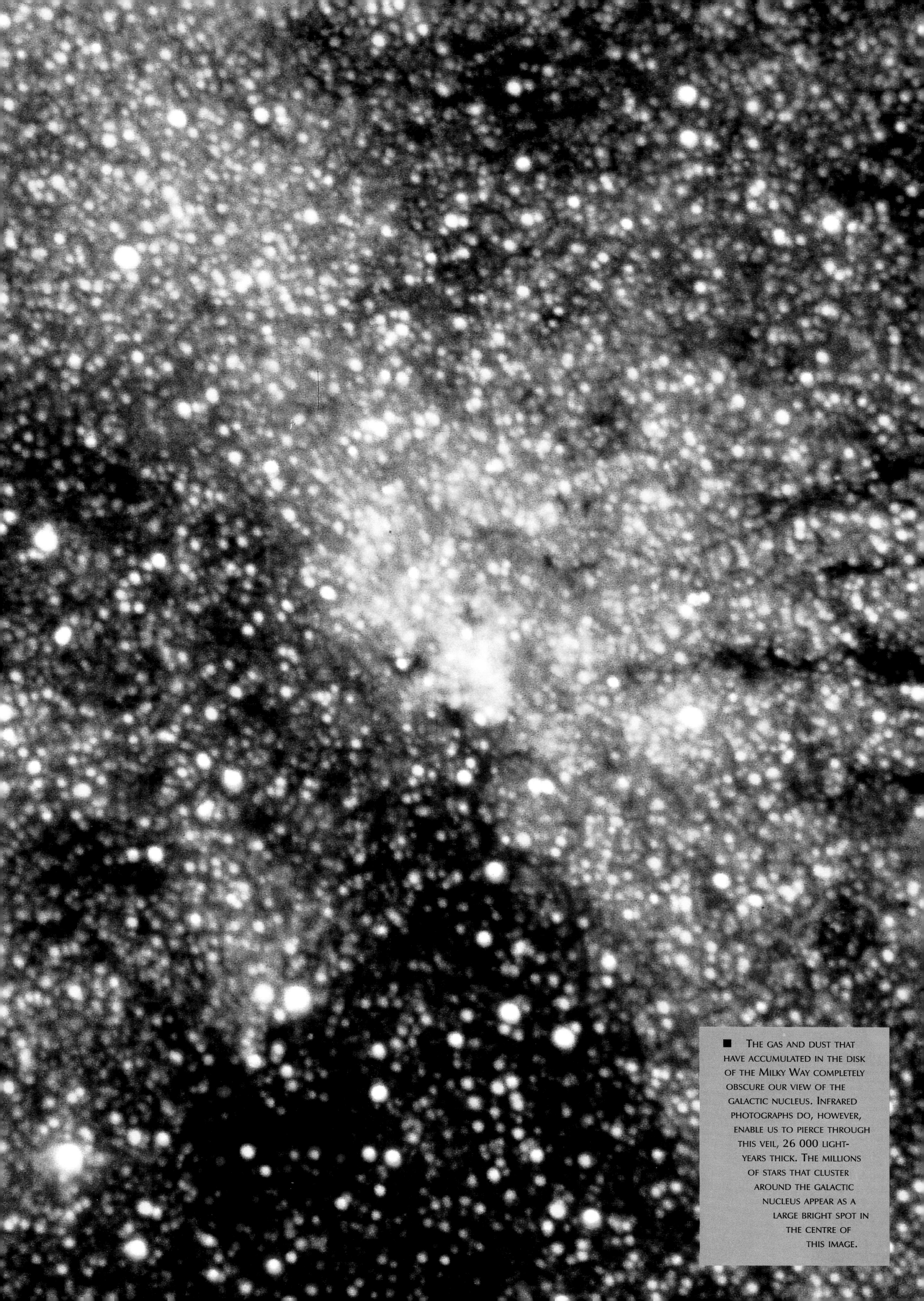

■ THE GAS AND DUST THAT
HAVE ACCUMULATED IN THE DISK
OF THE MILKY WAY COMPLETELY
OBSCURE OUR VIEW OF THE
GALACTIC NUCLEUS. INFRARED
PHOTOGRAPHS DO, HOWEVER,
ENABLE US TO PIERCE THROUGH
THIS VEIL, 26 000 LIGHT-
YEARS THICK. THE MILLIONS
OF STARS THAT CLUSTER
AROUND THE GALACTIC
NUCLEUS APPEAR AS A
LARGE BRIGHT SPOT IN
THE CENTRE OF
THIS IMAGE.

emits! Specialists say that the interstellar absorption in this direction reaches thirty magnitudes.

THE GALACTIC CENTRE

Yet astronomers have known for a long time that this unique point in the Galaxy does truly exist. And for a very good reason: all the stars in the sky are orbiting around it! It was measurements of the motions of the stars, at different distances from the Sun, beginning in the early years of the 20th century, that first revealed the overall shape of the Galaxy, and then allowed its centre to be located with a high degree of accuracy. Although the vast spiral disk, in which the Sun is located, extends for about 80 000 light-years, and the central bulge for around 18 000 light-years, the nucleus is restricted to a sphere of stars that is just 2000 light-years in diameter. Even though no one has ever seen the galactic centre, it is not very difficult to get an idea of its general appearance. All the nearby major spiral galaxies that are either seen face-on, or at least in almost full view, show the same sort of nucleus; a tiny, compact, and brilliant point of light, in sharp contrast to the relative faintness of their vast stellar disks.

It has been possible to estimate the mass of the mysterious and fascinating galactic centre from the velocities of the stars that orbit around it. The body, or the group of bodies, that hold sway over the immense structure of the Galaxy is itself a cosmic leviathan. Currently, astronomers estimate that 10 billion solar masses exist within a radius of 1000 light-years of the centre. This is an incredible stellar density, with nothing in common with the value for the region of the disk near the Sun. In our immediate vicinity there may be just about two stars in a cube, 10 light-years on a side. In the galactic nucleus, the same volume probably contains more than 10 000 stars! Seen from a planet orbiting one of the stars in this region, the sky must be ablaze: a blinding cloud of thousands of stars as bright as Venus, Jupiter or Sirius in our own sky. Nothing of the outside universe would be visible, apart from these stars.

The galactic centre proper lies at the very heart of the nucleus. As they studied other galaxies, astronomers soon realized that its features must be even more extreme than those of the nucleus. The stellar density, in particular, seemed to increase towards the centre at a totally incredible rate. But it was extremely frustrating for the researchers: there was this extraordinary object relatively close at hand, so to speak, yet it remained invisible, and proved as difficult to study as the nuclei of galaxies, 100 or 1000 times as far away. How could they pierce the shroud of stars and interstellar dust that hides the galactic centre from our view? With the help of

■ THE TENS OF BILLIONS OF STARS THAT ARE CONCENTRATED IN THE GALACTIC BULGE ARE INVISIBLE AT RADIO WAVELENGTHS. IN THIS IMAGE, NARROW, DENSE STREAMERS OF GAS ARE ESCAPING AT HIGH VELOCITY FROM THE EXTREMELY CHAOTIC CENTRE OF THE GALAXY.

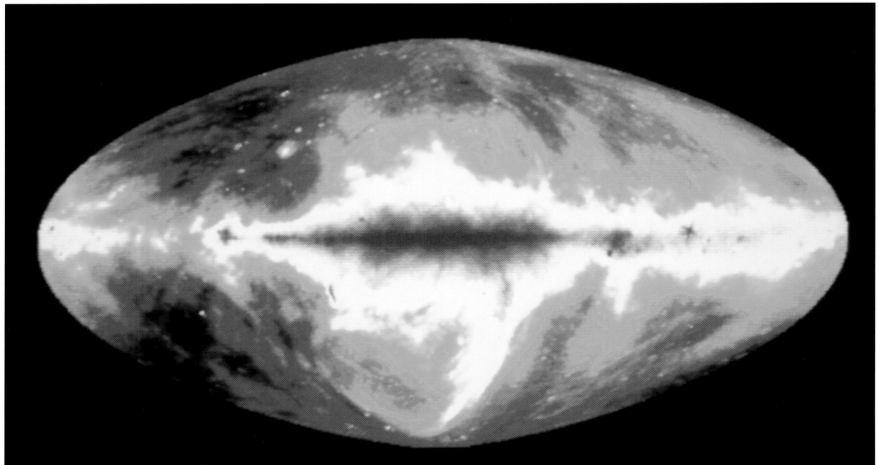

■ THIS IMAGE, OBTAINED AT RADIO WAVELENGTHS, SHOWS THE WHOLE OF THE DISK OF THE MILKY WAY. THE GALACTIC NUCLEUS IS IN THE CENTRE. UNLIKE LIGHT, RADIO WAVES ARE UNAFFECTED BY THE INTERVENING GAS AND DUST.

advances in technology, astronomers have finally partially overcome the difficulties.

It was through the use of radio telescopes that they initially gained access to the central regions of the Milky Way. Stars, in fact, emit very little at radio wavelengths. At wavelengths of 1, 10 or 20 cm, the Galaxy is practically transparent. Only interstellar gas is observed in this region of the electromagnetic spectrum. It was by mapping the gaseous clouds of the disk that researchers were able to discover the spiral structure, some forty years ago, and finally, about twenty years ago, actually penetrate right into the centre. Much more recently, major optical telescopes have been fitted with electronic cameras sensitive to infrared radiation. In this spectral region, between 1 and 100 microns, the veil that prevents us from seeing the galactic centre is far less effective. Finally, space observatories fitted with detectors sensitive to X- and gamma-rays are also able to pick up radiation emitted by the enigmatic central object. Astronomers soon realized that the nucleus, even though it was quite small when compared with the whole Milky Way, hid an object that was far smaller and had quite disconcerting properties. This object is so small, that as astronomical instrumentation has progressed, and the resolution of radio and infrared images has increased, astronomers have repeatedly been forced to to revise its dimensions downward. By studying the motions of the gas within the nucleus, researchers first identified a region some 30 light-years across, where millions of stars appeared to be crammed together, whirling at ridiculously high rates around the true centre, an extraordinary point in space that repeatedly seemed to elude them. At the beginning of the 1980s, the 27 aerials of the giant VLA radio interferometer, in New Mexico, obtained the first images of the centre at wavelengths between 1 and 21 cm. All that appeared on these strange images, after much computer-processing, was a chaotic nebula, consisting of trails of gas in the form of a distorted spiral. This

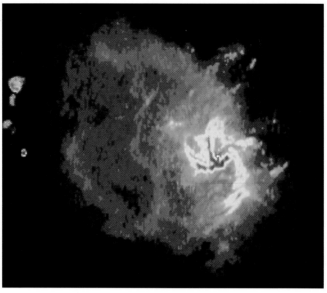

■ THIS GENERAL VIEW OF SGR A* WAS OBTAINED AT A WAVELENGTH OF 21 CM, WITH THE ARRAY OF 27 RADIO TELESCOPES THAT FORMS THE VLA IN NEW MEXICO. THE GALACTIC CENTRE, INVISIBLE ON THIS SCALE, IS AT THE CENTRE OF THE SPIRAL-SHAPED NEBULOSITY.

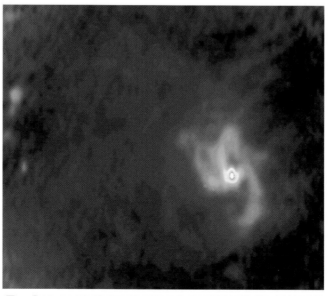

■ STILL CLOSER TO THE CENTRE: THE GALACTIC NUCLEUS, SEEN WITH THE VLA AT A WAVELENGTH OF 2 CM. AT THE CENTRE OF THE SPIRAL STRUCTURE OF SGR A*, THERE IS A TINY YELLOW AND RED SPOT THAT PROBABLY HIDES A BLACK HOLE, WITH A MASS EXCEEDING 1 MILLION SOLAR MASSES.

nebula, which measures no more than 10 light-years across, seems to be slowly rotating around an extremely bright, point-like source, that lies at its precise centre. With the location of this source, called Sgr A* by astronomers, they had finally reached their goal: it is the true galactic centre.

SGR A*: THE GALACTIC CENTRE

Shortly after this, in the 1990s, infrared telescopes detected hundreds of supergiant stars, lying close to that point, and forming the most formidable concentration of stars in the whole Galaxy. These stars, which are comparable to Rigel or Deneb, are hardly visible on the infrared images, but in reality are 100 000 times as bright as the Sun. To date, 300 have been found within a 1.5-light-year radius of Sgr A*, and this cluster, which is known as IRS 16, is probably the only part that can be detected. An enormous number of fainter stars remain undetectable, hidden by the 26 000 light-years of obscuration. The stellar density in this region is one million times that found in the vicinity of the Sun. Astronomers cautiously advance an explanation for the incredible number of supergiant stars at the galactic centre. They suspect that in this region of the Galaxy the stars are extremely close together, and their separation is no more than 100 billion kilometres. Every 10 000 years, on average, this close proximity must cause two stars to collide. It is probably the resulting fusion that provokes the formation of giant bodies, which are undoubtedly the most massive stars in the whole Galaxy.

The sky, seen from there, must be absolutely ablaze, a blinding, evenly distributed canopy of light, ripped apart here and there by several hundred objects of utterly unbearable brilliance, each the equivalent in mass to 10 to 100 Suns! Swept along by the swirling maelstrom around the galactic centre, which they orbit in just a few years, these stars must form constantly changing, ephemeral constellations, repeatedly distorted by their headlong flight, and by supernova explosions.

But what are these stars actually orbiting? Sgr A* is practically

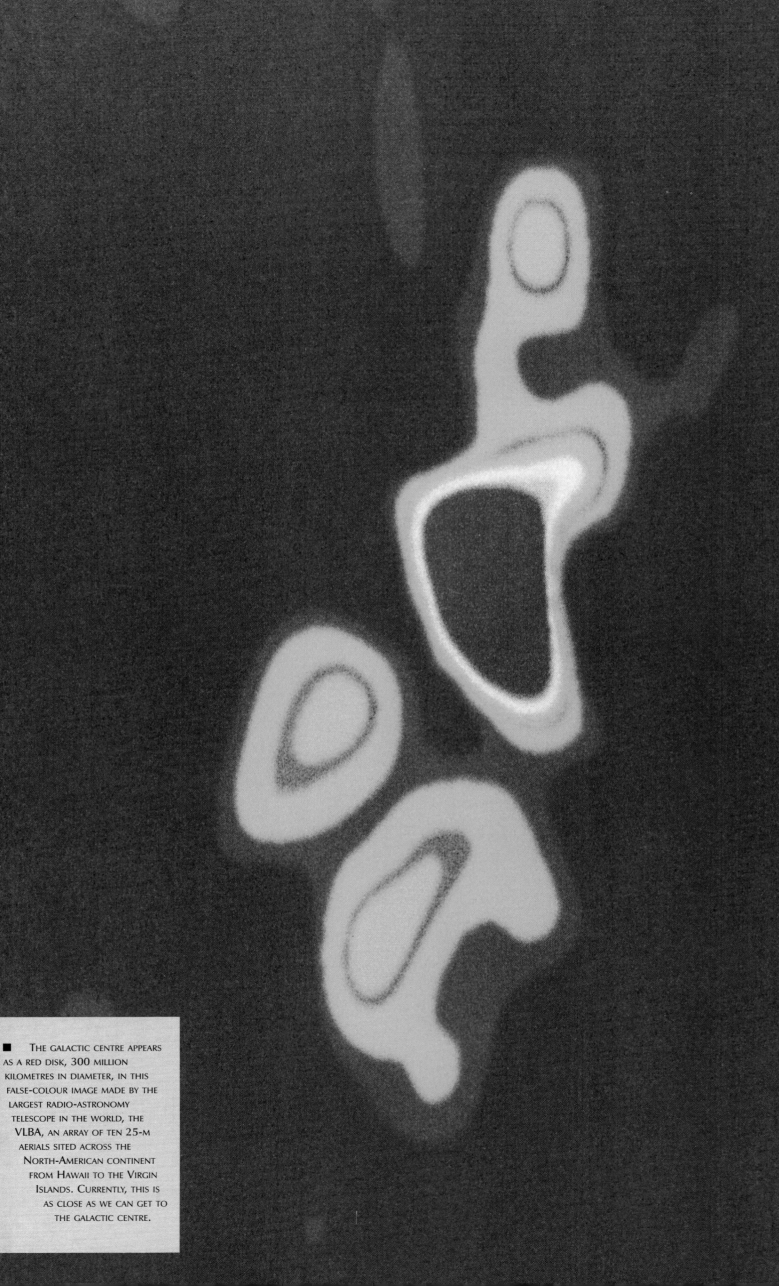

■ THE GALACTIC CENTRE APPEARS
AS A RED DISK, 300 MILLION
KILOMETRES IN DIAMETER, IN THIS
FALSE-COLOUR IMAGE MADE BY THE
LARGEST RADIO-ASTRONOMY
TELESCOPE IN THE WORLD, THE
VLBA, AN ARRAY OF TEN 25-M
AERIALS SITED ACROSS THE
NORTH-AMERICAN CONTINENT
FROM HAWAII TO THE VIRGIN
ISLANDS. CURRENTLY, THIS IS
AS CLOSE AS WE CAN GET TO
THE GALACTIC CENTRE.

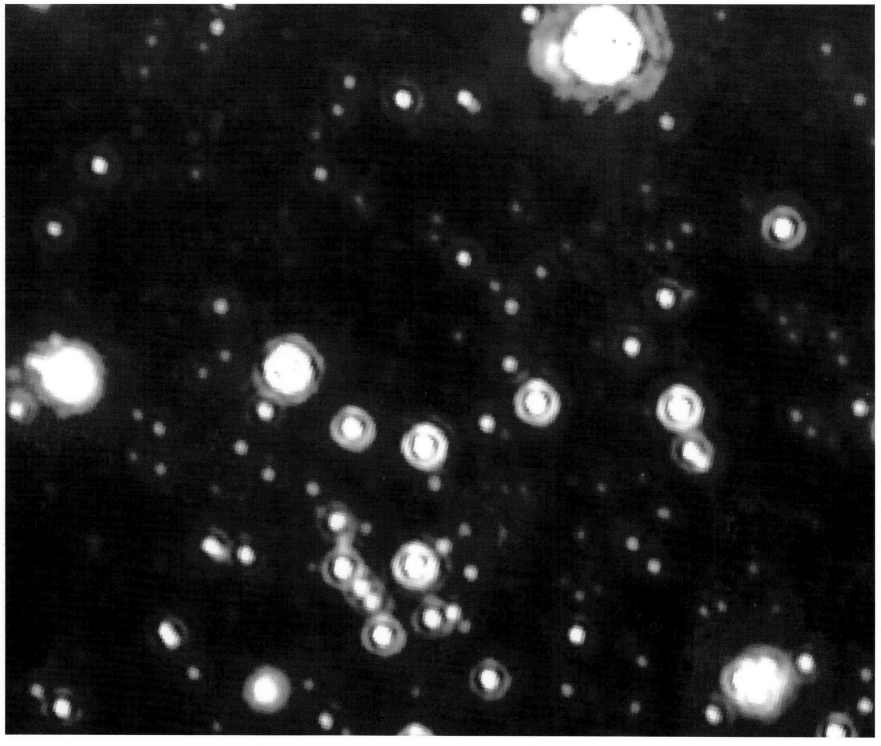

invisible to telescopes, with the powerful radio source appearing as nothing more than a faint, fuzzy point on infrared images. Currently, the object can be examined directly only with radio telescopes, and indirectly by measuring the influence that it exerts on the streamers of gas and the thousands of stars that surround it. The distribution of masses in the galactic centre has long intrigued researchers. In total, the equivalent of 2 million solar masses is in orbit around Sgr A*, but the mass of the central object itself must be more than 1 million solar masses. It remained to discover what this strange radio source was actually like: to measure its true size, which was something that, unlike its mass, it had refused to divulge, no matter how powerful an instrument had been used to examine it.

In 1992, some time after it had been commissioned, the largest radio interferometer on Earth, the VLBA, a network of ten telescopes installed right across the North-American continent, was turned onto the central point for the first time.

This impressive instrument produced an astonishing image of Sgr A*. The body, around which the whole Galaxy is orbiting, was finally revealed as a tiny elliptical spot, measuring no more than 300 million kilometres across, i.e. about twice the Earth–Sun distance. It cannot be a star: the most massive stars known are highly unstable, and can reach 100 solar masses only with great difficulty. An extremely tight cluster of stars? Such an idea appears difficult to accept: it would be impossible to collect several tens of thousands of Denebs or Rigels into such a small volume. Gravitational perturbations would soon cause all the stars to crash into one another and produce a general conflagration...

A BLACK HOLE AT THE CENTRE OF THE GALAXY

According to researchers, only one type of body as massive as Sgr A* could exist within such a small volume: a black hole. It is an absolute stellar monster that is hiding at the centre. This invisible object probably originally formed through the collapse of

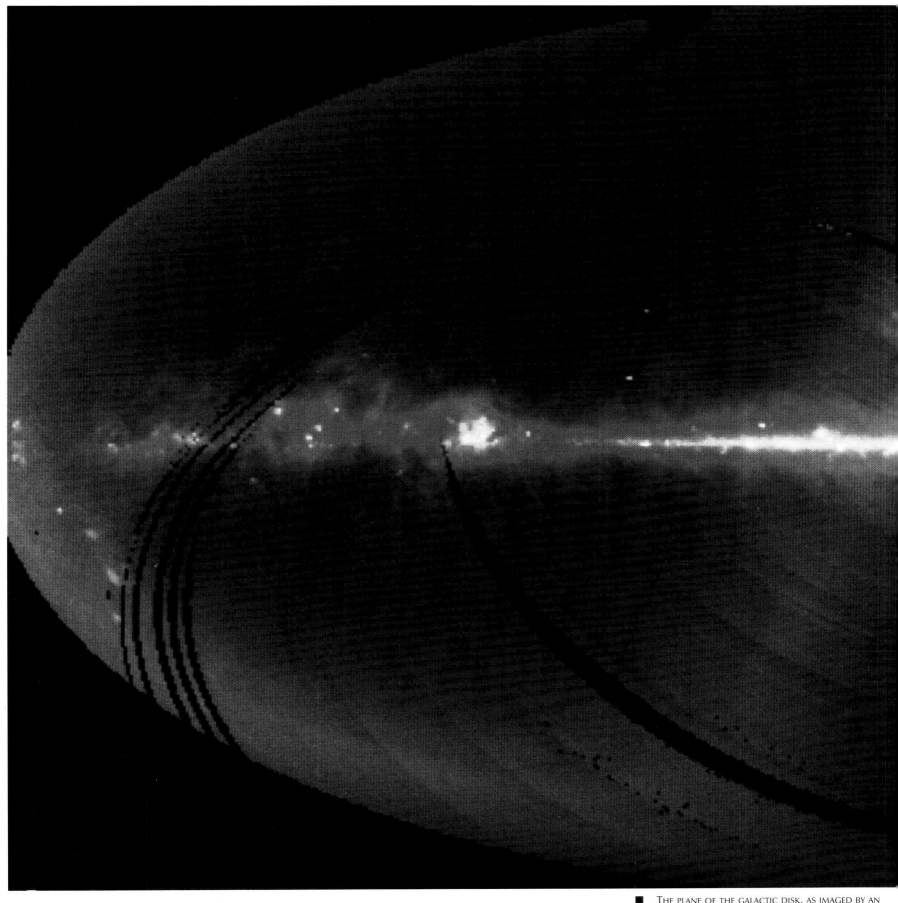

a supergiant star. In the incredibly crowded region around the galactic centre, stellar collisions often occur, at least on a cosmic time-scale, and the black hole must have attracted first one star, then a second, and a third... Over the course of time, this black hole's mass increased, and its gravitational field grew. The theory of relativity does not forbid these bodies from growing indefinitely. According to the astrophysicists, the smallest black holes form naturally, with the collapse of the cores of supernovae, which are themselves the final stage in the evolution of supergiant stars. Such black holes have a mass that lies between 3 and 10 solar masses, but are confined – as an unknown state of matter – in a volume that is less than 3 km in diameter. So far, no one has ever seen the immediate vicinity of a black hole, but, in principle, astronomers have access right up to its boundary, which is known as the event horizon. This event horizon is the boundary at which the black hole's gravitational field just allows light to escape. No physicist has the slightest idea of what happens beyond that horizon. In the case of the galactic centre, the influence of the black hole, which is proportional to its mass, should extend out to a distance of 5 million kilometres, i.e. 16 light-seconds.

In one or two decades from now, when telescopes are sufficiently powerful to allow researchers to study a region that is so distant and so small, perhaps we shall finally discover the true

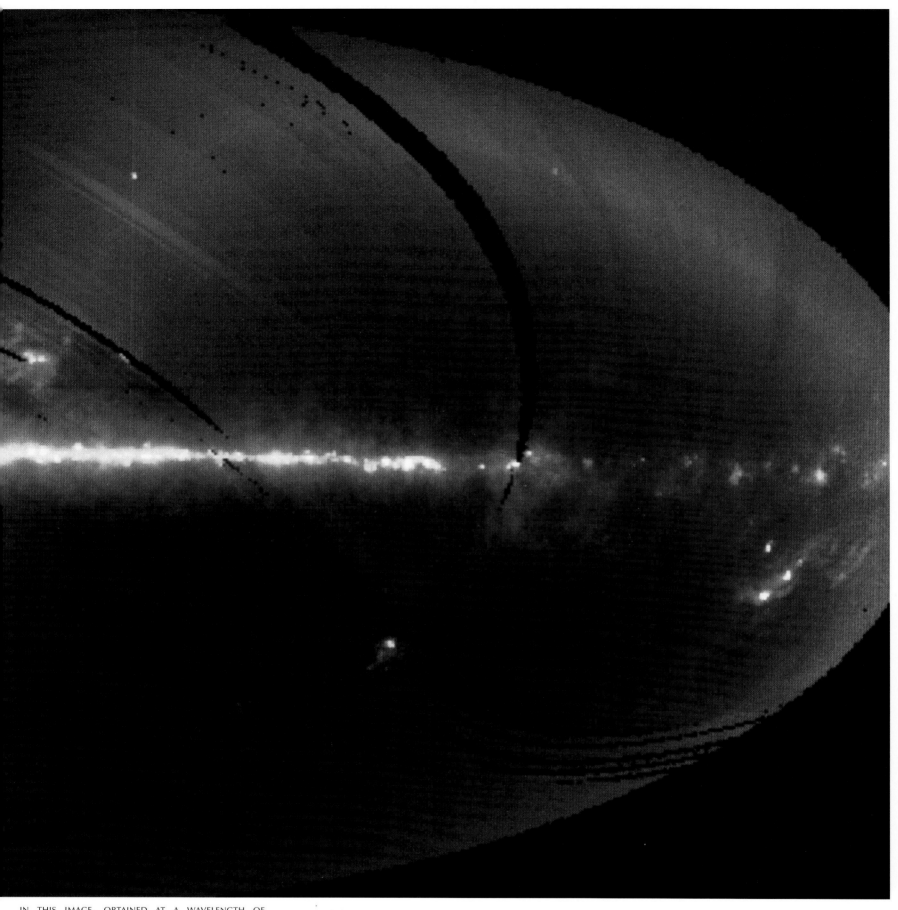

IN THIS IMAGE, OBTAINED AT A WAVELENGTH OF
100 MICRONS.

nature of Sgr A*. Astronomers suspect that they are
dealing with a disk of searingly hot gas that is orbiting a central,
invisible point, which is forever hidden behind its event horizon.
From time to time, streamers of gas, heated to several million
degrees, fall at nearly 300 000 km/s towards the event horizon,
where they disappear, giving rise to violent gamma-ray emissions
as they do so. The disk is probably mainly fed by the surges of gas
that are ejected from the hundreds of supergiant stars that
surround it, but also by entire stars, that are drawn near, torn
apart, and finally devoured by the intense gravitational field of the
monstrous black hole that is hidden in the centre of the Galaxy. ■

87

A sea of galaxies

■ THE GALAXY M 83, IN THE CONSTELLATION OF HYDRA. OUR OWN GALAXY, THE MILKY WAY, PROBABLY CLOSELY RESEMBLES THIS SPLENDID BARRED SPIRAL, WHICH LIES AT A DISTANCE OF SOME 12 MILLION LIGHT-YEARS, AND CONTAINS MORE THAN 100 BILLION STARS. IN THIS TRUE-COLOUR PHOTOGRAPH, THE DISK OF M 83 APPEARS BLUISH, BECAUSE IT CONSISTS OF MILLIONS OF VERY YOUNG STARS, WHILE THE YELLOWISH BULGE MAINLY CONTAINS OLD STARS.

■   M 100 IS ONE OF THE MOST MASSIVE SPIRAL GALAXIES KNOWN. IT LIES AT A
DISTANCE OF SLIGHTLY MORE THAN 50 MILLION LIGHT-YEARS, IN THE CONSTELLATION OF
VIRGO.

The universe is populated by galaxies, out to the very farthest reaches (both in space and time), visible to the most powerful telescopes. They are the largest organized structures in the universe. No form of larger-scale mechanism exists that might otherwise create still larger objects. On photographs taken with the largest telescopes, galaxies seem to be as numerous and fragile as snowflakes in a blizzard. But each one of the 'snowflakes' actually consists of stars. Tens, hundreds, and sometimes even thousands of billions of stars endlessly whirl around inside these ghostly shapes, which range from young, graceful spirals to old, massive ellipticals. Galaxies are true universes in themselves: at the very beginning of time they became concentrations of primordial hydrogen and helium. Carried along by the cosmological expansion, they recycle this material indefinitely, each shaping, in its own way and at its own pace, the gas in the nebulae into stars that are born, expand, and die within its confines. But how many galaxies are there in the sky, and how far away can we see them? Four of them, including the Milky Way, may be clearly seen with the naked eye. In the Northern Hemisphere, the gauzy spiral M 31 in Andromeda

appears, by a happy effect of perspective, to lie near the clouds of the Cygnus and Cepheus spiral arms. This pale, elongated glimmer of light resembles our own Galaxy seen from afar, and lies at a distance of 2.5 million light-years, i.e. approximately 25 billion billion kilometres. This distance, which seems incomprehensible, is a mere stone's throw on a cosmological scale. In the Southern Hemisphere, two other galaxies that are visible to the naked eye are more spectacular than M 31, because they are much closer. They look like clouds or pieces torn from the Milky Way. The Large Magellanic Cloud is a small galaxy that lies at about 170 000 light-years, while the Small Magellanic Cloud is about 220 000 light-years away. Beyond these three, galaxies are too far away, and their apparent brightness is too low, for them to be seen by the naked eye. With binoculars, a careful observer would certainly be able to see more than one hundred, and an experienced amateur astronomer could, over the course of a lifetime, find several thousand with a small telescope. At one time, astronomers had an ambition to catalogue all the galaxies. Charles Messier, Louis XV's astronomer, discovered about sixty, and at the end of the 19th century, the Danish astronomer Johan Dreyer recorded

■ THE SPECTACULAR, BARRED SPIRAL
GALAXY NGC 2442 LIES IN THE
CONSTELLATION OF VOLANS. THIS
PHOTOGRAPH WAS OBTAINED WITH
THE 3.9-M ANGLO-AUSTRALIAN
TELESCOPE AT SIDING SPRING
OBSERVATORY IN AUSTRALIA. IT
SHOWS THE MARKED BAND OF
INTERSTELLAR DUST THAT FOLLOWS
THE SPIRAL ARMS, AND WHICH
ALSO INDICATES THE LOCATION
OF THE DENSITY WAVE.
MILLIONS OF STARS ARE
CURRENTLY BEING BORN
IN NGC 2442.

nearly 10 000, after three decades of observations. At the turn of the century, astronomers – who were quite ignorant of the true nature of the small, indistinct patches of light that they were finding on the sky – still had no idea of the utter impossibility of the task that they had set themselves. Nevertheless, as the power of telescopes increased, so too did the number of galaxies... James Keeler, who studied them with the Crossley 91-cm telescope at Lick Observatory, estimated in 1900 that their total number was slightly more than 100 000.

Between the 1950s and the 1980s, astronomers at the observatories at Palomar in the Northern Hemisphere, and La Silla and Siding Spring in the Southern, commissioned Schmidt telescopes (a form of powerful, wide-field camera), with the aim of mapping the whole sky. The sensitive plates that were obtained were so rich that scanners, linked to powerful computers running shape-recognition software, had to be used to make a census of galaxies. Several tens of millions had been recorded on the photographic plates. Since then, no further census of galaxies has been attempted. Researchers content themselves by probing specific areas of the sky with their more powerful telescopes, and apply the results of their measurements to the whole sky, using statistical methods. According to the most recent measurements, made with the 10-m Keck Telescope on Hawaii, and with the Hubble Space Telescope from orbit, the number of galaxies that would be visible, nowadays, down the limit of detectability, which is thirtieth magnitude, exceeds 50 billion galaxies.

These tens of billions of galaxies do, however, fall into just four main categories, which were recognized early in the 20th century: the spirals, ellipticals, lenticulars, and irregulars.

SPIRAL GALAXIES

The most spectacular of all these bodies are undoubtedly the spirals, the prototype of which is our own Milky Way. Like it, spirals are galaxies that have both large sizes and large masses. The least massive contain some ten billion stars. The Milky

■ THE GIANT, FACE-ON, BARRED SPIRAL GALAXY, NGC 1365, PHOTOGRAPHED FROM SIDING SPRING OBSERVATORY. THIS GALAXY IN FORNAX IS PARTICULARLY PROLIFIC, AND CONTAINS MILLIONS OF YOUNG, BLUE SUPERGIANT STARS, WHOSE COLOUR MAY BE SEEN IN THE STRIKING SPIRAL STRUCTURE.

Way, with its diameter of 80 000 light-years and an overall mass equivalent to 200 billion Suns at least, is a very respectable spiral. The largest of its siblings, however, are even more impressive. Our neighbour, M 31 in Andromeda, is at least twice as massive, and the largest spirals known, NGC 6872, NGC 1961 and UGC 2885, are perhaps ten times as massive, and ten times as large. The absolute record is currently held by the galaxy NGC 6872, in the constellation of Pavo, 200 million light-years away. It is 800 000 light-years in diameter.

Like human beings, all spirals resemble one another, but like them, no two are really identical. Even comparing thousands of galaxies as recorded photographically by large telescopes, astronomers always find subtle differences in their morphology, some individuality, that may be explained by their history, which is unique to each one of them, and consists of bursts of star formation, the explosion of supernovae, and even of stormy encounters with other galaxies.

A spiral galaxy is a vast stellar disk, thicker at the centre (the bulge), within which lies a tiny, brilliant, and extremely massive nucleus. The bulge consists of low- and medium-mass stars that are several billion years old. It is the disk that is the site of the magnificent spiral structure, which is literally illuminated by millions of brilliant young stars. The disk of spirals always has the same general appearance: first a broad region that separates two spiral arms, a zone that is populated by older, fainter stars, very similar to the Sun; then the inner edge of an arm, like a charcoal drawing, a sort of spiral in negative, often as dark as space itself, created by the dark nebulae. Then there is a necklace of ionization nebulae, which, largely consisting of hydrogen, tint the galactic arms with delicate touches of pastel pink. Finally, the spiral arms proper, long and drawn-out, thicker and thinner in places, which coil two or three time around the galaxy. These consist solely of several tens of thousands of supergiant stars, which give rise to their brilliance and their characteristic sky-blue colour. The spiral arms move more slowly than the general flow of stars, so to an external observer the spiral structure would appear to rotate in the opposite direction! This is a case where the physical reality appears contrary to common-sense, but which is clearly demonstrated by the fine structure of the arms. First comes the interstellar gas, cold and dark, which condenses inside the arms, then the bright nebulae, and their retinue of young stars that illuminate them. So stars enter a spiral arm from the inner side, cross it, and leave on the outside.

All similar and yet different, spiral galaxies are classified into eight major groups, proposed by the American astronomer Edwin Hubble in 1925, and which we shall use here (rather than the later, but less aesthetically pleasing, and more complex classification devised by Gérard de Vaucouleurs). First there are the Sa galaxies, followed by the Sb, Sc, and Sd, each step representing a decrease in the prominence of the bulge relative to the disk. In the Sc spirals, for example, the bulge is practically invisible, represented solely by the nucleus. Their enormous disks, by contrast, have extremely luminous spiral arms, which are often multiple, irregular, and fragmented. When the spiral structure springs from a greatly elongated bulge, the spirals are known as 'barred', and classified SBa, SBb, SBc, and SBd. We still do not know to which class we should assign our own galaxy. Some researchers believe that it is almost identical to M 31, the majestic Sb spiral in Andromeda, or the beautiful galaxy NGC 2997. Others maintain that the Milky Way is a barred spiral, type SBb, which may resemble M 83 in Hydra. Astronomers have been trying to understand the origins of the spiral structure in galaxies for decades. Despite the galaxies' regularity and almost mathematical beauty, they have yet to solve the mystery.

For a long time it was wrongly believed that a single dynamical process could explain all facets of the phenomenon, despite its diversity. For example, some spirals have just one or two arms, which wrap round the disk several times. Others, probably including the Milky Way, have three or four arms. One key factor in understanding spiral structure was discovered following the statistical study of several thousand galaxies. There are no dwarf spiral galaxies. The initial mass of

the galaxy therefore seems to be a factor. Recently, thanks to numerical simulations on supercomputers, theoreticians have been surprised to discover that, in a rotating disk consisting of tens of billions of stars, a spiral structure appears spontaneously, and then slowly dissipates after a few hundred million years. Once a spiral structure has arisen, however, it needs to persist, which the early simulations did not succeed in reproducing. Researchers finally discovered that to restart the process, or maintain it, the subtle balance of the galaxy's gravitational field needs to be slightly perturbed from time to time – by an encounter with another galaxy, for example. Examples are legion, in the universe, of pairs or groups of galaxies where spiral arms are distorted, swollen, or seem to be entwined as a result of the gravitational attraction from a neighbouring galaxy. This explanation accounts for simple spiral structure. But numerous galaxies have disks that are broken up into innumerable short segments of arms, which form an almost filamentary spiral structure of astonishing complexity and delicacy. In this latter case, we are probably dealing with a sequence of exploding supernovae, whose shockwaves compress the surrounding interstellar medium, precipitating the formation of young stars, which explode in their turn. It is the most luminous of these stars, the blue supergiants, which, over the course of time, trace out the spiral structure of the galaxies, like a spark following a trail of gunpowder. The differential rotation of the galactic disk completes the process; the lines of young stars are naturally distorted into a spiral.

ELLIPTICAL GALAXIES

Unlike the spiral galaxies, ellipticals display no specific structure, and no spectacular shape. They have a perfectly simple and regular appearance, and resemble smooth ellipsoids of revolution. Put more simply, they resemble large, almost spherical snowballs, or like more or less elongated eggs. Edwin Hubble, in his famous catalogue of galaxies, classified them according to their degree of flattening, from E0 to E7. The former appear spherical, while the latter are the most elongated.

Elliptical galaxies include both the smallest and the largest galaxies. Dwarf

■ THE GALAXY NGC 3992 LIES IN THE CONSTELLATION OF URSA MAJOR. ITS MAGNIFICENT SPIRAL STRUCTURE IS AMAZINGLY SYMMETRICAL. SMALL BLUISH CLUMPS ARE JUST VISIBLE ALONG THE GALAXY'S SPIRAL ARMS. THESE ARE REGIONS OF INTENSE STAR FORMATION.

A CLOSE-UP OF THE GALAXY M 100 IN VIRGO. THIS PHOTOGRAPH, TAKEN WITH THE HUBBLE SPACE TELESCOPE, SHOWS WITH EXTRAORDINARY CLARITY THE COMPLEX SPIRAL STRUCTURE OF THE GALAXY. NUMEROUS SUPERGIANT STARS, ABOUT 100 000 TIMES AS BRIGHT AS THE SUN, ARE SCATTERED THROUGHOUT THE GALACTIC DISK. FIVE SUPERNOVAE HAVE BEEN DETECTED IN THIS GALAXY DURING THE 20TH CENTURY.

NGC 253 IS A SPIRAL GALAXY
IN THE CONSTELLATION OF SCULPTOR.
IT LIES AT A DISTANCE OF SOME
12 MILLION LIGHT-YEARS. SEEN
ALMOST IN PROFILE, THIS SPIRAL HAS
A DISK THAT IS PARTICULARLY RICH
IN GAS AND DUST. THE SMALL RED
SPOTS THAT MAY BE SEEN IN THIS
PHOTOGRAPH, SCATTERED
AROUND THE DISK OF
NGC 253, ARE NEBULAE,
COMPARABLE WITH THE
ORION NEBULA IN THE
MILKY WAY.

ellipticals, such as the Draco and Leo 2 galaxies, contain less than one million stars, and their brightness is about 100 000th that of the Milky Way. In contrast, supergiant ellipticals are utterly outstanding objects. Some of them, more than ten times as bright as the Milky Way, contain an incredible number of stars. These extraordinary objects, which may consist of 10 000 billion stars, are the largest galaxies known. The stellar population of elliptical galaxies, from the smallest to the largest, is remarkably homogeneous. Practically devoid of interstellar gas, they do not contain any nebulae, and therefore do not form any new stars. Because young, hot, massive, and very bright stars are completely absent from these galaxies – if they did exist, they would be easy to discover, either on photographs or through their spectra – astronomers conclude that stellar formation ceased several billion years ago in ellipticals. They are therefore very ancient objects, containing some hundreds of thousands to thousands of billions of low-mass stars, mainly red giants, and red and white dwarfs.

■ NGC 6822 is one of the satellites of the Milky Way. It is a small irregular galaxy, at a distance of less than 2 million light-years. The brightest stars in the galaxy are clearly visible in this photograph, which was taken with the 3.9-m Anglo-Australian Telescope at Siding Spring.

Lenticular and irregular galaxies

Lenticular galaxies (S0 in Hubble's classification) have characteristics common to both ellipticals and spirals. Like ellipticals, they have a significance bulge, consisting of old stars. Like spirals, they have retained a disk of stars. But this disk is practically devoid of gas and young stars, and shows no trace of spiral structure, which is proof that the lenticulars, like the ellipticals, are no longer the site of star formation.

There remain the irregulars. As their name indicates, these galaxies do not display any recognizable geometrical shape, nor any particular symmetry in their organisation, although they may sometimes exhibit an embryo bulge and disk, and thus look like a half-finished sketch of a spiral. Like the latter, however, irregulars contain both old and young stars. Very rich in gas – nebulae may represent 10 to 30% of their mass – irregulars contain numerous blue supergiants that are just a few million years old.

Irregulars are – apart from the dwarf ellipticals – the smallest and least massive galaxies. NGC 6822 in Sagittarius, for example, is just 6000 light-years in diameter, with a mass of no more than 400 million solar masses. Others, like the Small Magellanic Cloud, reach two billion solar masses, and the most massive irregulars probably exceed ten billion solar masses. This is relatively modest when compared with a giant spiral like the Milky Way, which, at the lowest estimate, contains 200 billion stars, or possibly two or three times that number.

The distribution and history of galaxies

What is the distribution of these four major classes of galaxies in the vicinity of the Milky Way? If we include only the brightest objects, with at least ten billion stars, out to a few tens of million light-years, we find that more than 60% are spiral galaxies, about 20% lenticulars, 15% ellipticals, and less than 5% irregulars. However, such a count favours luminous galaxies over dwarfs. In fact, astronomers have recently discovered that dwarf galaxies are probably the most numerous, even though their masses are negligible with respect to the large spiral and elliptical galaxies.

It still remains for us to discover the origin of Hubble's major classes of galaxy. Some people, including Hubble himself, originally thought that they represented a chronological sequence, with ellipticals gradually evolving into spirals by way of lenticulars. Nowadays, the specialists are absolutely certain that the four Hubble classes of galaxies do not represent evolutionary stages. As proof of this, very old stars typical of ellipticals are found in certain regions of irregular galaxies, as well as in the bulges of spirals.

Scientists have thus nowadays arrived at the conclusion that most galaxies were born at more or less the same time, at an era that corresponds to the oldest stars found within them. According to this scenario, it would have been the initial conditions under which they arose that transformed some of them into spirals and others into ellipticals.

But when and how were galaxies born? We do not know. According to the Big-Bang theory, matter in the universe, which was originally searingly hot and homogeneous, cooled, fragmented, and condensed little by little, over the course of time. Some 12 to 15 billions years ago, the matter must have formed into vaguely spherical or ellipsoidal clumps: the protogalaxies. Under the influence of their own gravity, these vast condensations of gas must have slowly collapsed in on themselves, simultaneously acquiring a slow rotation. Centrifugal forces tend to flatten such a rotating cloud of gas, and cause it to assume the shape of a disk, thickened towards the axis of rotation. This is the shape found in spirals. Once the density of the gas in the protogalaxy was sufficiently high, a second sequence of fragmentation and of collapse was initiated, this time giving rise to stars. In the case of ellipticals,

some theoreticians believe that the rotation of the gaseous body was slower and that star formation started sooner, during the course of the protogalaxy's collapse. According to them, this explains why, in an elliptical galaxy, the stars are distributed regularly, with the stellar density increasing in step as one gets closer to the centre.

But there remain various areas of doubt. Why, for example, have both ellipticals and lenticulars entirely recycled their interstellar gas into stars? Why, on the other hand, are spirals and irregulars, 12 or 15 billion years after their formation, still creating new stars at the rate of ten, one hundred, or even a thousand every year?

Slowly the idea has come to be accepted that galaxies, over the course of billions of years, have all evolved, have all had their own particular history. To discover, once and for all, this hidden history of galaxies, astronomers had no choice; they had to study the universe on a far greater spatial and temporal scale. Precise observations of each of the individual bricks – even though there might be tens of billions of them – would be in vain. They needed to examine the architecture of the universe itself, on the scale of billions of light-years, to understand the evolution of galaxies. ▪

GALAXY, WHICH CONCEALS A MASSIVE BLACK HOLE AT ITS
CENTRE, CONTAINS SEVERAL THOUSAND BILLION STARS,
MOST OF WHICH ARE VERY ANCIENT.

The architecture of the universe

■ Around 50 million light-years away from the Milky Way, the Virgo Cluster contains about 3000 galaxies. In this photograph, only the very centre of this vast swarm is visible, dominated by the two giant elliptical galaxies, M 84 and M 86. Each of these galaxies contains around 10 000 billion stars. On the right-hand side, a spiral galaxy appears to be distorted by the powerful gravitational attraction of its two massive neighbours.

INTERSTELLAR DUST CLOUDS IN THE DISK OF THE ANDROMEDA GALAXY APPEAR SILHOUETTED AGAINST ITS BRILLIANT CENTRAL BULGE. ASTRONOMERS SUSPECT THAT M 31 HARBOURS A GIANT BLACK HOLE AT ITS CENTRE.

Although galaxies are the building blocks of the universe, they are not evenly distributed in space, but have a natural tendency to occur in groups. Our own galaxy, the Milky Way, for example, is not isolated in the universe, but forms the centre of attraction for a few much smaller galaxies. At what is a mere stone's throw, on a cosmic scale naturally, its imposing neighbour the Andromeda Galaxy, itself reigns over a second small extragalactic court. The Milky Way and M 31 in Andromeda are the two principal members of a small, gravitationally bound group, known as the Local Group. Seen from outside, from a great distance (if that were possible), the Local Group would present a wonderful sight to an interstellar voyager. The latter would first of all see two brilliant spiral galaxies poised in a sort of frozen ballet, floating in space, and separated by the considerable distance of 2.5 million light-years. Around these two giants there would be a cloud of luminous motes, some easily visible, small but bright, and others almost imperceptible, minute, and translucent, almost hidden in the reassuring arms of the spirals. Several dozen galaxies in all, which, trapped by the powerful gravitational forces of the Milky Way and M 31, constantly

revolve around them, completing their orbits is several hundred million years. The Large and the Small Magellanic Clouds are the principal satellites of the Milky Way, and lie at distances of 170 000 and 220 000 light-years, respectively. Their masses are 10 billion solar masses for the first, and 2 billion for the second. But far smaller galaxies also exist extremely close to us. They are so inconspicuous that they were not discovered until the 1950s: Leo 1 and Leo 2, or the Draco system, which contains no more than 100 000 stars.

In Andromeda, M 31 is a sort of super-Milky Way. It is larger – the disk of this giant galaxy is more than 150 000 light-years in diameter – and it is also more massive. Containing perhaps 1000 billion stars, M 31 is probably as massive as all the other galaxies in the Local Group combined. This giant is an Sb spiral, and resembles our own Galaxy. It is sufficiently close to us for a number of its individual stars to be visible, in particular the red giants, blue supergiants, and variable stars. It is a remarkable astrophysical laboratory for research workers who are able to see (with hindsight), what happens in the Milky Way. At the distance of the Andromeda Galaxy, in fact, blue supergiants – like Rigel or Deneb in our Galaxy – are easy to

102

detect, because they have apparent magnitudes as high as magnitude 17. Stars like Vega in Lyra, Arcturus in Boötes, or Aldebaran in Taurus at about magnitude 25, are far more difficult to detect in the dense fog of stars in the galaxy. As for stars similar to the Sun, at magnitude 29, they remain invisible today, lost in the silvery arms of the beautiful spiral, even if we try to observe them with the Hubble Space Telescope.

LIFE IN THE LOCAL GROUP

Astronomers suspect that the nucleus of the Andromeda Galaxy – like the centre of the Milky Way and many other galaxies in the universe – shelters a giant black hole. Although the one in the Milky Way has a mass of around 3 million solar masses (according to recent estimates), the black hole in the Andromeda Galaxy would appear to be as much as 70 million solar masses. The hypothetical black hole in M 31 – just like the one in the Milky Way – is not directly accessible to our optical instruments. It is, in fact, a minute body – at least on a galactic scale. How can we see a point that is less than one light-minute across, at a distance of 2.5 million light-years?

The Andromeda Galaxy is under constant surveillance by amateur and professional astronomers. Hundreds of variable stars, most of which are pulsating red and yellow supergiants, identical to Mira Ceti and to the Cepheids in the Milky Way, are scattered throughout its disk. Occasionally, extremely close pairs of stars, one of which is a white dwarf, exchange streams of material. This gas builds up on the surface of the white dwarf and suddenly ignites. For a short space of time, i.e. a few days or weeks, the luminosity of the binary increases by a factor of 10 000. This phenomenon, known as a nova, is observed in the Milky Way, not just in the Andromeda Galaxy, in which several dozen novae have been discovered over the course of the 20th century. Above all, however, in such a massive galaxy, with so much interstellar gas and populated by thousands of

blue supergiants, we can expect one of them to explode sooner or later. Just such an event occurred in 1885, and we know that if a bright supernova should appear in the Andromeda Galaxy, it could be followed with the naked eye from Earth.

M 32 and NGC 205 are the two principal satellites of the giant spiral. They are very old dwarf ellipticals, which have not given birth to any new stars for at least five billion years. Their masses – for dwarf ellipticals – are fairly significant: 3 billion solar masses for M 32, and more than 10 billion solar masses for NGC 205. Finally, another delicate spiral galaxy, M 33 in Triangulum, hangs in space not far from M 31. The disk of this Sc-type galaxy exhibits broad, chaotic arms, sparkling with thousands of blue supergiants, and banded with nebulosity, which in places has been tangled by the shockwaves from old supernovae. At the end of the northern arm of M 33 there is a gigantic nebula, unlike anything in our own Galaxy. This nebula, NGC 604, gave birth to several hundreds of blue supergiants some millions of years ago. It will probably be the site of a supernova explosion in the near future. M 33 lies at a distance of 2.7 million light-years, slightly farther than M 31 in Andromeda. On extremely clear autumn nights, some experienced amateur astronomers with particularly well-adapted vision are able to see it with the naked eye.

Most astronomers are convinced that the Local Group includes about one hundred mini-galaxies, rather than the thirty-odd bodies known today. First, because certain dwarfs are almost undetectable. The last dwarf galaxies, the Carina and Sagittarius dwarfs, were only discovered in the Local Group in 1974 and 1976, respectively. Second, because other galaxies, although intrinsically bright, remain invisible, hidden behind the clouds of stars and nebulae in the arms of the Milky Way. This was proved by the astounding discovery made by Butler Burton and Renée Kraan-Korteweg in August 1994 with the 25-m radio

■ THE GLOBULAR CLUSTER 47 TUCANAE, ONE OF THE 125 CLUSTERS THAT ORBIT IN THE HALO OF OUR OWN GALAXY, THE MILKY WAY. THE GIANT ELLIPTICAL GALAXY M 87 IS SURROUNDED BY MORE THAN 15 000 GLOBULAR CLUSTERS. EACH CONTAINS BETWEEN 100 000 AND 1 000 000 STARS.

telescope at Dwingeloo in the Netherlands. Far beyond the clouds of stars in the Milky Way, in the constellation of Cassiopeia, these Dutch astronomers discovered, one after the other, two galaxies very close to the Local Group, which had gone unnoticed until then. The first of these newcomers is impressive: Dw 1 is a barred spiral about 60 000 light-years in diameter and with a mass approaching 60 billion solar masses. Dw 2 is half the size and has about one tenth the mass. Both objects lie at distances of more than 10 million light-years.

The few dozen galaxies in the Local Group are gathered into an almost spherical volume of space that is slightly more than 5 million light-years in diameter. Beyond that, no matter in which direction they look, astronomers do not observe any galaxies at all, dwarf or giant, until about 10 million light-years. This is the distance at which the next group of galaxies is found, in Sculptor. About half as rich as the Local Group, the Sculptor Group includes some spectacular galaxies such as NGC 55, NGC 247, NGC 253, and NGC 300. All these galaxies – which are seen at various angles, the first in profile, and the last face-on – are close enough for astronomers to be able to observe their most luminous stars individually.

In the region immediately surrounding the Milky Way, there are about 50 groupings similar to the Local and Sculptor Groups. Each is dominated by one, two or three large galaxies. The average distance between them is 10–20 million light-years.

CLUSTERS: SWARMS OF GALAXIES

But the organization of the universe does not stop there. Structures that are vastly larger than groups, known as clusters of galaxies, are visible all over the sky. Although their size is hardly any larger than that of the groups, the number of galaxies within them is enormous. The closest of these swarms of galaxies lies in the constellation of Virgo, at a distance of 50 million light-years. The Virgo Cluster contains close to 3000

galaxies in a volume of space that is about 6 million light-years in diameter. In particular, it includes the galaxies M 99 and M 100, which seem to enclose a central, more concentrated zone that is populated solely by lenticulars and giant elliptical galaxies. At the centre of the Virgo Cluster, the galaxies are extremely close to one another. In some cases their separation is only slightly greater than their diameter. The heart of the Virgo Cluster is occupied by M 84, M 86, and M 87, three giant ellipticals. The last is one of the most massive galaxies known. Ten times the luminosity of the Milky Way, it contains more than 10 000 billion stars. Its halo, which extends 200 000 light-years from the centre, is populated by more than 15 000 globular clusters, whereas our Galaxy, for comparison, has about one hundredth of that number. Finally, M 87 is not like other galaxies. Its extraordinarily bright nucleus is the source of a luminous jet, consisting of a series of brilliant clumps of material, all precisely aligned with one another. This straight, yet turbulent 'flame' shoots out for about 6000 light-years, and appears to contain more energy than several billion stars. The origin of M 87's jet is to be sought in the paroxysms of activity

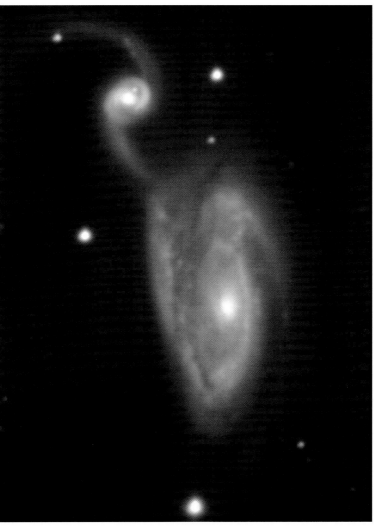

in the galaxy's nucleus, which releases an enormous amount of energy. The nature of its emission is quite well understood. It is produced by free electrons that are moving at a velocity close to that of light in a powerful magnetic field. The energies involved are extraordinary. To eject this plasma over such a distance, and doubtless for several million years, the nucleus of M 87 must produce energy about ten billion times more than that released by the Sun. To try to solve this enigma, two of the most powerful telescopes, the Hubble Space Telescope and the VLBA, were turned on M 87 in 1994. These instruments were able to show that the jet originates less than one light-year from the centre of the galaxy, in a rapidly rotating disk of material. Currently, theoreticians know of only one engine capable of generating so much energy in such a

■ THE SPECTACULAR ENCOUNTER BETWEEN TWO GALAXIES, NGC 5394 AND NGC 5395, IN THE CONSTELLATION OF CANES VENATICI. THESE TWO SPIRALS LIE AT A DISTANCE OF NEARLY 150 MILLION LIGHT-YEARS FROM THE MILKY WAY. THE RESEMBLANCE BETWEEN THIS PAIR AND THE DOUBLE GALAXY M 51, SHOWN OPPOSITE, IS EXTREMELY STRIKING.

■ The M 51 pair in Canes Venatici lie at a distance of less than 15 million light-years. After having passed less than 100 000 light-years from one another, the two galaxies are slowly separating. The arms of the large spiral appear to have been slightly deformed by the encounter, while the small galaxy, an old spiral, is nothing more than a chaotic mixture of stars and nebulae.

small volume – a black hole. The one in M 87 is probably enormous. Hidden behind its event horizon, this body attracts interstellar gas – or rather stars – from the giant galaxy. Part of the material that spirals in towards the black hole is expelled at 10 000 km/s, confined by the lines of force of an intense magnetic field that coincides with the disk's axis of rotation. It is this stellar plasma that forms M 87's jet. According to the latest models, the central black hole in M 87 amounts to some 3 billion solar masses, i.e. 1000 times that in the

■ THIS CLOSE-UP OF THE GIANT ELLIPTICAL GALAXY M 87 (SEE P.104) WAS OBTAINED WITH THE HUBBLE SPACE TELESCOPE. A JET OF PLASMA, 6000 LIGHT-YEARS LONG, IS BEING EXPELLED FROM THE HIGHLY LUMINOUS NUCLEUS OF THE GALAXY, WHERE A MASSIVE BLACK HOLE IS PROBABLY LURKING.

giant black hole hidden in the heart of the Milky Way. We shall see later why a massive galaxy like M 87 is at the centre of the Virgo Cluster.

Despite its 3000 galaxies, specialists do not consider the Virgo Cluster to be a particularly unusual cluster as far as the universe as a whole is concerned. Astronomers now know of thousands of clusters that are far richer, such as Abell 370, Abell 1689, or Abell 1656. Often extremely distant, many of

these clusters remained invisible until the 1960s. Abell 1656, in the constellation of Coma Berenices, is a notable exception. Located at 'only' 350 million light-years, it is the closest to us. This relative proximity means that it is far easier to study. With this famous cluster, which astronomers also know as the Coma Cluster, the scale changes dramatically. More than 10 000 galaxies are crowded together, only a few tens of million of light-years apart, into a sphere less than 10 million light-years across. The Coma Cluster is dominated by two of the most luminous and massive galaxies known: NGC 4878

■ THE GALAXY NGC 4151, IN THE CONSTELLATION OF CANES VENATICI, LIKE M 87, IS EMITTING A JET OF PLASMA AT SEVERAL THOUSAND KM/S. IN THE CASE OF NGC 4151, THE JET IS POINTING TOWARDS US, AND APPEARS AS AN EXTREMELY BRILLIANT POINT OF LIGHT.

and NGC 4889, monsters that probably each contain more than 10 000 billion stars, and which are almost certainly more massive than M 87 in the Virgo Cluster. The overall mass of the Coma Cluster exceeds one million billion solar masses.

The Coma Cluster has a remarkably spherical and homogeneous structure. The distribution of its galaxies increases smoothly from its outer regions towards the centre. Astronomers have not found even the smallest spiral in its central regions, which are populated solely by elliptical and lenticular galaxies. The same factor had already been noted in the Virgo Cluster, and soon proved to be universal. Spirals are always absent from regions of the universe that contain large numbers of giant elliptical galaxies. Study of clusters of galaxies has allowed this puzzle over galactic selection to be resolved, and at the same time, to understand finally the physical processes to which the four great Hubble classes correspond. In a cluster that is very rich in galaxies, the latter may lie incredibly close to one another. At the centre of the Virgo or Coma Cluster, they are separated by distances that are three to six times their diameter, or even less. Like planets around stars, the small galaxies revolve around the larger ones on elliptical orbits, with periods of several hundred million years. During the course of their slow revolution, the spirals undergo strong gravitational perturbations. Stars are relatively insensitive to these, but on the other hand, they have a considerable influence on the large unstable masses of gas found in nebulae. Scientists believe that over the course of time spirals lose their gas, which escapes into space. Denuded of the principal material from which stars are made, these galaxies age without creating new generations of stars, and from being young spirals, they become old lenticulars. The discovery, at the beginning of the 1980s, that clusters of giant galaxies are embedded in a halo of extremely tenuous diffuse gas – amounting to about one atom per cubic decimetre – which may have come from these old spirals that have turned into lenticulars, tends to support this theory.

In clusters, however, galaxies are not content with just passing one another at a distance, leaving the gas from their nebulae behind in space. Sometimes they approach one another too closely, and driven by powerful gravitational forces, slowly collide with one another. It is an amazing spectacle to see two spirals, each amounting to 100 billion solar masses, crash headlong into one another in almost infinitely slow motion. The collision, which does, in fact, occur at a few hundred kilometres per second, takes several hundred million years. The shock would probably not be perceptible to a hypothetical inhabitant of one of the two spirals. The distances between the stars are so enormous, relative to their size, that the hundreds of billions of stars can hurtle towards one another and pass by without any collisions taking place. On the other hand, the dynamics of the two galaxies is dramatically altered. The disks warp and become deformed, the spiral arms break up, the nebulae collide and suddenly heat up, giving rise to new swarms of stars. Finally, when the two galaxies, each on its own orbit, separate, they fling out into intergalactic space

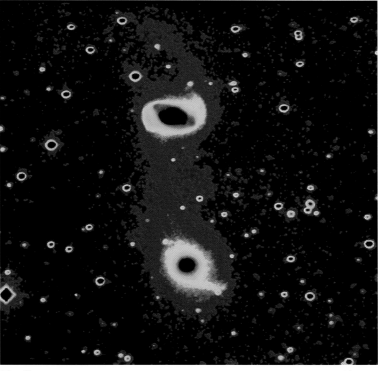

■ THE STATELY DANCE OF GALAXIES LASTS HUNDREDS OF MILLIONS OF YEARS. HERE, A VAST ARC OF STARS, EXTENDING FOR HUNDREDS OF MILLIONS OF LIGHT-YEARS, MARKS THE TURBULENT ENCOUNTER BETWEEN THE TWO SPIRAL GALAXIES OF THE PAIR KNOWN AS ARP 104, AND WHICH OCCURRED LONG AGO.

immense bridges of stars that stretch between them. These arcs of stars, hurled out into space, extend for hundreds of thousands of light-years before slowly spreading out and disappearing. The two galaxies, stripped of gas, but glowing with millions of blue supergiant stars, slowly recover their equilibrium as they separate. They have a natural tendency to reform with a spiral structure, in which a bar sometimes develops, which helps them to recover their dynamical stability.

Some astronomers believe that new galaxies are born from the ribbons of stars that are lost to space. In France, Félix Mirabel and Pierre-Alain Duc have discovered a dwarf galaxy forming in the immense trail of stars ejected by Arp 105, a pair of galaxies in the cluster Abell 1185. The apparent calm in our own Local Group, with its quiescent galaxies separated by several hundred thousand light-years is deceptive. The Milky Way, for example, has probably devoured dwarf galaxies in the distant past, and it seems that in due course (in 20 billion years) the orbits of the Magellanic Clouds will cause them to fall into the Milky Way and merge with it. Even more spectacular will be the collision of the Milky Way with the giant Andromeda Galaxy. The two galaxies are currently closing at the rate of 100 km/s, and will pass through one another in less than 3 billion years.

Although high-velocity encounters allow each galaxy to emerge more-or-less free from the unwanted gravitational effects, it also happens that two galaxies may slowly fuse together. When astronomers discovered this amazing phenomenon, they realized that they had simultaneously

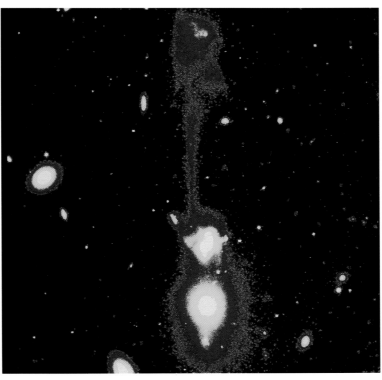

■ THE STRANGE SYSTEM ARP 105 LIES IN THE ABELL 1185 CLUSTER OF GALAXIES. AT THE BOTTOM OF THE IMAGE, A GIANT ELLIPTICAL PREPARES TO ENGULF A SPIRAL, WHICH, PERTURBED BY THE GRAVITATIONAL FIELD OF ITS MASSIVE NEIGHBOUR, EMITS A GIGANTIC JET OF STARS AND GAS OUT INTO SPACE.

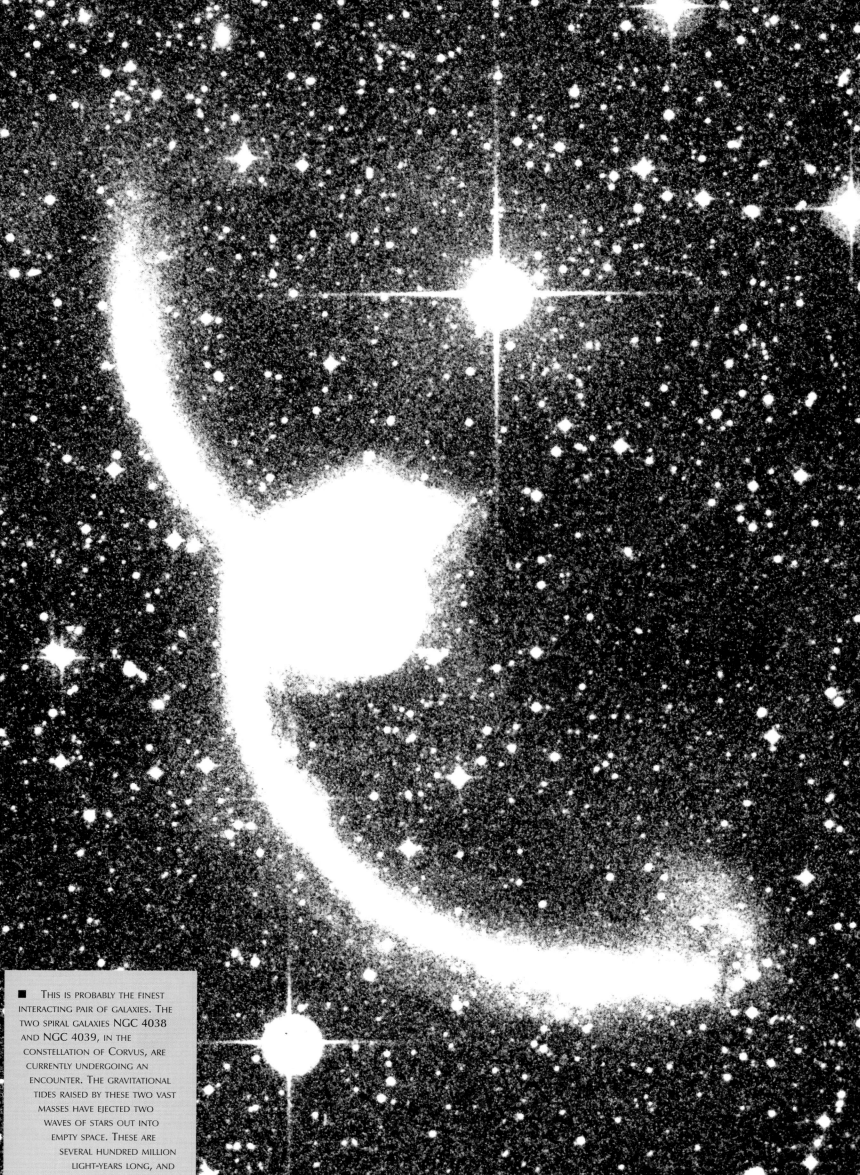

■ THIS IS PROBABLY THE FINEST
INTERACTING PAIR OF GALAXIES. THE
TWO SPIRAL GALAXIES NGC 4038
AND NGC 4039, IN THE
CONSTELLATION OF CORVUS, ARE
CURRENTLY UNDERGOING AN
ENCOUNTER. THE GRAVITATIONAL
TIDES RAISED BY THESE TWO VAST
MASSES HAVE EJECTED TWO
WAVES OF STARS OUT INTO
EMPTY SPACE. THESE ARE
SEVERAL HUNDRED MILLION
LIGHT-YEARS LONG, AND
WILL SOON DISPERSE.

found the hitherto mysterious origin of giant elliptical galaxies. These monstrous objects swallow up whole galaxies one by one. The closest of these galactic cannibals is 12 million light-years away, in the constellation of Centaurus. NGC 5128 is a giant elliptical, with a mass that probably exceeds 1000 billion solar masses. It is a strange, two-faced object. Overall, it resembles a normal giant elliptical. But the centre of this galaxy is surrounded by clouds of interstellar dust, bright nebulae, and young stars – just the sort of characteristics that one finds in the disks of spiral galaxies. And for good reason: astronomers have realized that the galaxy in Centaurus is a giant elliptical, caught as it calmly engulfs a small spiral. In a few hundred million years, the gas from the dead spiral will have completely disappeared, and NGC 5128 will not retain any traces of its incredible astronomical meal. It is probably similar cannibalism, but of another order of magnitude, that produced the enormous M 87 galaxy in Virgo. Its enormous mass, and its halo that is abnormally rich in stars, suggest that over the last 10 billion years numerous galaxies have slowly merged to create this giant elliptical. The gas from the spirals that have been torn apart by its intense gravitational field, may have fallen towards its centre of mass, and might at the same time explain the existence of M 87's giant black hole.

■ NGC 5128 IS AN ELLIPTICAL GALAXY SLOWLY FUSING WITH A SPIRAL. THE DISK OF THE SPIRAL, ITS NEBULAE, AND ITS CLOUDS OF INTERSTELLAR DUST ARE STILL VISIBLE AROUND THE CENTRE OF THE ELLIPTICAL.

THE FATE OF GALAXIES AND THE STRUCTURE OF THE UNIVERSE

The puzzling organization at the centre of clusters of galaxies, which are rich in giant ellipticals yet devoid of spirals, now becomes clear. Once again, it is through studying clusters, whose compact structure greatly promotes galactic interactions, that has allowed astronomers to sketch a general outline of the history of the galaxies. Broadly speaking, these were almost all born at roughly the same time, let us say a few hundred

■ THE GALAXY CLUSTER ABELL 1060, AS PHOTOGRAPHED FROM SIDING SPRING. IN THIS RATHER SPARSE CLUSTER, ALL TYPES OF GALAXIES STILL EXIST TOGETHER: ELLIPTICAL, LENTICULAR, AND SPIRAL.

million years after the Big Bang. Contrary to a widely held, and extremely intuitive, view, they did not subsequently evolve individually, retaining their initial form and moving apart in accordance with the general expansion of the universe. Such a view only holds on a very large scale, much greater than that of clusters of galaxies. In fact, galaxies have a very eventful history, with multiple encounters and mergers, and even changes of identity, inside clusters of galaxies. At present no one knows what is in store for the Milky Way. Will it continue to have an elegant spiral structure after it has passed right through the Andromeda Galaxy? Or, on the other hand, will the two spirals merge to form a single giant elliptical, in which several thousand billion stars will be concentrated in an indescribably disordered fashion?

Most galaxies probably belong to groups or clusters of galaxies like the Virgo and Coma Clusters. We do not know if galaxies exist that are completely isolated in intergalactic space. Astronomers have surveyed around 10 000 clusters, out to a distance of slightly more than 1 billion light-years. It is on this scale of billions of light-years that the true architecture of the universe begins to be seen. When we consider things on such a vast scale, individual galaxies can no longer be seen, of course, and even clusters appear no more than faint, indistinct points of light. Once again, even on this enormous scale, we observe a tendency towards clustering. Systems that are even larger than clusters of galaxies exist, and contain tens of clusters. Our own Local Group is gravitationally influenced by the Virgo Cluster, which is slowly pulling us towards it. This Local Supercluster, which was discovered in 1950 by the astronomer Gérard de Vaucouleurs, is an elongated swarm, centred on the Virgo Cluster, that contains more than 100 000 galaxies, spread over a

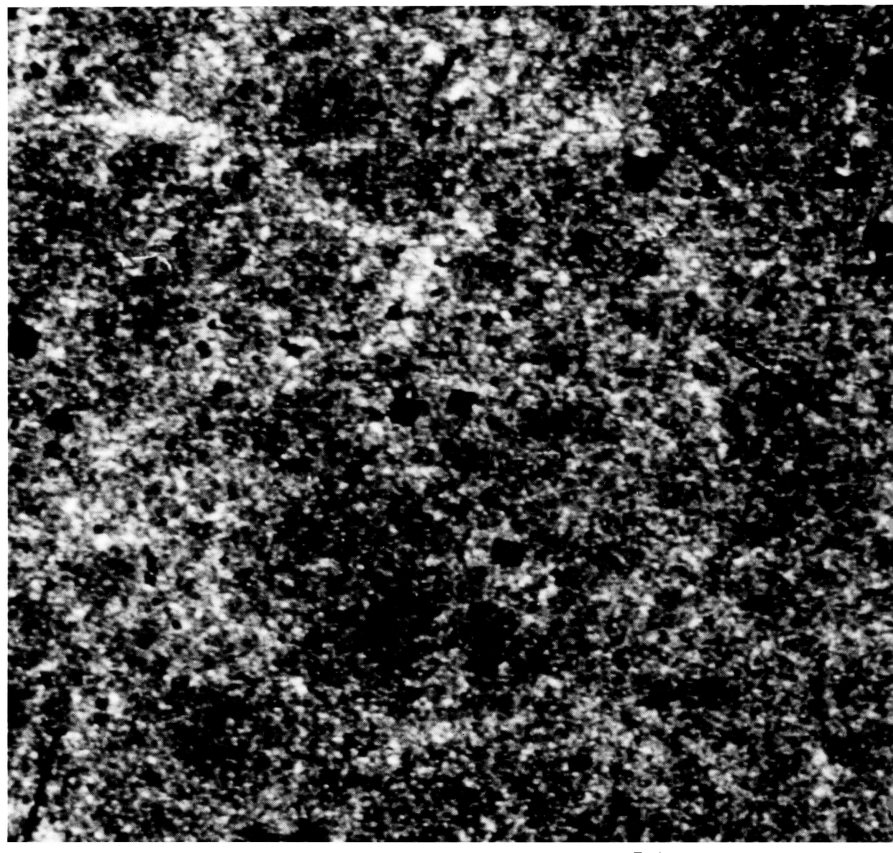

distance of more than 50 million light-years.

On a still larger scale, the universe probably has a cellular structure. Obtaining evidence of this particular structure has required decades of patient observation, which only met success at the end of the 1980s. The difficulty is explained when we realize that all the bodies studied by astronomers – stars, nebulae, and galaxies – appear projected against the dome of the sky. How can we reconstruct the third dimension, the volume of space, and ensure each object is in its correct place and at its correct distance? In addition, the galaxies that astronomers study in carrying out their three-dimensional celestial cartography mostly lie at distances of millions of light-years, and appear as very faint glimmers of light. Recording their images and their spectra requires whole nights of observation, even with giant telescopes equipped

with electronic cameras. Without new spectro-graphs that are capable of simultaneously measuring the distances of a hundred different galaxies, we would still have no idea of what the cosmic landscape looks like on a very large scale. To gain some idea of what it is like, imagine that the universe is like a bubble bath. Each bubble is one of the cells discovered by the researchers. The walls of the bubbles consist of clusters of galaxies. The densest superclusters resemble the lines where two bubbles intersect. Each one of these cosmic bubbles, most of whose volume is empty space, measures about 300 million light-years across. The organization of the universe appears to stop here. ▪

WHICH GALAXIES OCCUR IN CLUSTERS IS OBVIOUS, AND
THE OVERALL CELLULAR STRUCTURE OF THE UNIVERSE IS
BEGINNING TO APPEAR.

The Big Bang, the story of the Universe

■ THE OLDEST GALAXIES KNOWN ARE ABOUT 12 BILLION YEARS OLD. DURING THEIR EVENTFUL EXISTENCE, THEY ENCOUNTER OTHER GALAXIES, WITH WHICH THEY FORM TEMPORARY OR MORE LASTING LINKS. HERE, IN THE CONSTELLATION OF PEGASUS, ARE FOUR OF THE FIVE GALAXIES IN THE SPECTACULAR STEPHAN'S QUINTET. THE SPIRAL NGC 7320, BOTTOM, IS IN FRONT OF THE GROUP. IT LIES AT A DISTANCE OF 50 MILLION LIGHT-YEARS, AS AGAINST 350 MILLION LIGHT-YEARS FOR THE OTHER THREE GALAXIES.

THE SUPERGIANT ELLIPTICAL GALAXY IC 4051, IN THE COMA CLUSTER, IS SURROUNDED BY A THOUSAND GLOBULAR CLUSTERS, CLEARLY VISIBLE ON THIS IMAGE. TO ASTRONOMERS, THESE BODIES ARE VALUABLE DISTANCE INDICATORS.

Out in the depths of space, myriads of galaxies are lost in a limitless abyss, each with arms sparkling with hosts of stars, scattered with ephemeral lesser worlds, swarms of invisible planets and moons, all fleeting clouds of individual atoms... . But is this hierarchical structure of bodies that we see around us locally, and at this period of time, really representative of the architecture of the universe? Has matter always assumed the same forms – clusters, galaxies, nebulae, stars, and planets – as those that we see today? This extremely intuitive image of an eternal, unchanging universe, and one that Albert Einstein naturally referred to when he sketched out his first theoretical model of the universe, reflects the cosmological principle, one of the fundamental postulates of science: the universe has the same properties everywhere, and our region of space is representative of the universe as a whole. Extended to the concept of the universe's existence in time, this becomes the perfect cosmological principle: the universe is the same everywhere, and at all times.

But does the universe actually obey this simple ideal? Have stars been burning for an infinite time? Do galaxies appear immutable, out to the farthest distances that our telescopes can reach? We now know that the answer to such questions is 'No', and the discovery that the universe is evolving, that it has a history, will probably remain the most significant scientific and philosophical discovery of the millennium.

When observed with the world's largest telescopes, galaxies have the appearance of being captured in full flight against the background tapestry of the cosmos. Their distances from Earth – millions or hundreds of millions of light-years – does not allow us to detect the slightest movement in these island universes. Of course, astrophysicists are able to use their calculations to reveal the slow rotation of a giant galaxy, or the spectacular ballet performed by two spirals that are flinging streams of stars out towards one another. The rotation of individual galaxies, and their orbits within clusters where thousands of them are gathered together, hide an overall motion that is almost imperceptible, and which is yet of a far greater significance: the expansion of the universe.

As seen from Earth, and excluding our nearest neighbours in the Local Group, all the galaxies appear to be receding. This motion, which is not directly visible through a telescope, is

■ AT SOME 350 MILLION LIGHT-
YEARS FROM THE MILKY WAY LIES
THE GIANT COMA CLUSTER: AROUND
10 000 GALAXIES, WHICH, BECAUSE
OF THE EXPANSION OF THE
UNIVERSE, ARE MOVING AWAY FROM
US AT 7000 KM/S. THIS IMAGE OF
THE COMA CLUSTER, LIKE THAT
ON P.116, WAS OBTAINED WITH
THE SPACE TELESCOPE. IT
SHOWS THE FINE SPIRAL
D 216, AND, TO THE LEFT,
THE FAINT HALO AROUND
THE GIANT ELLIPTICAL
NGC 4881.

■ The Leo Cluster lies at
about 40 million light-years.
Centred on two giant
ellipticals, this low-density
cluster contains two fine
spirals, M 95 and M 96, seen
clearly at the bottom of the
image. The hundreds of stars
visible on this very wide-
field photograph are some
of the Sun's neighbours.
They belong to our
own galaxy, the
Milky Way.

revealed by spectral analysis of their light. If a source of light moves relative to an observer, the latter finds that the wavelength of the radiation that is detected has altered. This is the famous Doppler effect. When a source is approaching us (like the Andromeda Galaxy, for example), all the spectral lines are shifted towards the blue. Their wavelength is decreased. When the source is receding (like all the galaxies, except those in the Local Group), the wavelengths are increased, and the lines are shifted towards the red. Precise measurement of this spectral shift enables us to obtain an importance piece of astronomical information: the speed at which bodies are moving.

■ A CLOSER VIEW OF THE GALAXY M 96 IN LEO. DETAILS OF THE SPIRAL ARMS BEGIN TO APPEAR, IN PARTICULAR, SOME CLUSTERS OF YOUNG BLUE STARS. ON THE IMAGE BELOW, OBTAINED BY THE HUBBLE SPACE TELESCOPE, ONE OF M 96'S SPIRAL ARMS EXHIBITS THOUSANDS OF RED AND BLUE SUPERGIANTS.

AN EXPANDING UNIVERSE

It was, of course, by studying their spectra that astronomers discovered that galaxies were receding from us. No real significance was placed on this recession before Edwin Hubble carried out measurements at Mount Wilson Observatory and, in 1929, published the law that bears his name: all the galaxies are receding from us at a velocity that increases in proportion to their distance. The recession of the galaxies was immediately interpreted as a phenomenon of cosmological significance. Astronomers remembered that the theory of general relativity contained the seeds of a model in which the universe expanded. In fact, although Einstein himself set such a model aside, preferring, for aesthetic and philosophical reasons, to develop a static cosmological model, three theorists each found models

■ THIS FINAL CLOSE-UP OF M 96 SHOWS THE FRACTAL STRUCTURE OF THE UNIVERSE. BEHIND THE THOUSANDS OF STARS IN THE DISK OF M 96, NEW GALAXIES APPEAR. THIS IMAGE RESEMBLES THAT ON P.118, OPPOSITE, AND IF TELESCOPES WERE POWERFUL ENOUGH, WE CAN EASILY IMAGINE THAT THIS SEQUENCE COULD BE REPEATED TO INFINITY... .

of an expanding universe within the equations of relativity. These were Willem De Sitter in 1917, Aleksandr Friedmann in 1922 and, in particular, Georges Lemaître in 1927. According to relativity, the recession of the galaxies is simply a reflection of the dilation of space or, expressed more generally, of the expansion of the universe.

The Hubble law is one of the simplest laws of physics, and is written as $V = H_0 D$. In other words, the recession velocity of a galaxy is equal to the product of its distance times the famous Hubble Constant, H_0. Although all the galaxies appear to be receding from us, this does not imply that the Milky Way is at the centre of the universe. In fact, the observer's position is irrelevant. An astronomer would see exactly the same recession of the galaxies from any other galaxy whatsoever. H_0 is expressed in km per second per megaparsec (km/s/Mpc), a megaparsec being a unit of distance that is used by astrophysicists, and which represents a distance of about 3 million light-years.

Thanks to measurements made by the Hubble Space Telescope between 1993 and 1996, we now know that the Hubble Constant is – within observational error – about 75 km/s/Mpc. To put this in more concrete terms, the Virgo Cluster, with the thousands of galaxies that surround the mysterious giant elliptical M 87, is receding at a velocity of around 1200 km/s. Its distance, according the Hubble law, is thus around 50 million light-years. The Coma Cluster, one of the largest concentrations of galaxies currently known in the universe, is receding at 7000 km/s, so its distance exceeds 300 million light-years. As for the cluster Abell 370, in the constellation of Cetus, its recession velocity is more than 100 000 km/s. The Hubble law gives it a distance of more than 4 billion light-years.

These velocities, according to the relativistic interpretation of the observations, are not real, but apparent ones. According to Einstein's theory, it is the framework of space-time, its geometry, that is expanding, carrying with it the pockets of matter that form the galaxies. One can be scientifically rigorous and still state that these bodies are essentially at rest, but that they are receding from one another, simply because the scale of the universe is changing with time.

Nevertheless, astronomers soon said to themselves that if galaxies are receding from one another, they must once have been much closer. But how far back in time? It was Georges Lemaître who, in 1931, suggested that the cosmological expansion had a specific origin in time, by putting forward the theory – later popularized under the name of the Big Bang – that in the distant past, the entire universe was concentrated into a single point, which the Belgian mathematician called the primordial atom. Cosmologists nowadays prefer to call this point at which the universe originated – and which Lemaître envisaged as subsequently 'exploding' – the 'initial singularity'. This is a mathematical expression which takes account in a better way of the indeterminate nature of its initial state, which may, in fact be unknowable. According to the physicists, the universe becomes a reality an infinitely short time after the Big Bang, where its temperature and density tend

■ THE OLDEST STARS KNOWN HAVE BEEN FOUND IN GLOBULAR CLUSTERS, COLLECTIONS OF HUNDREDS OF THOUSANDS OF STARS, WHICH, ACCORDING TO SOME ASTRONOMERS, ARE THE OLDEST OBJECTS IN THE UNIVERSE. THIS IS THE HEART OF M 15, A GLOBULAR CLUSTER BELONGING TO OUR GALAXY, THE MILKY WAY.

to infinity. According to the Big-Bang theory, the universe was originally a flood of pure energy. The four fundamental forces – gravitation, electro-magnetism, and the weak and strong nuclear forces – were initially unified, but became decoupled. Then, as space expanded, matter – rigorously following Einstein's equations – appeared, at first searingly hot, but slowly cooling. As space expanded, the universe cooled and started to evolve. Atoms appeared, they condensed into gas, which gathered into stars and galaxies. The sole task of modern cosmology is to attempt to prove or disprove this overall theoretical scheme, and then to unearth its history, to date it, and to understand the most obscure and the most complex episodes.

This new scientific dogma – that the universe has a history – is nowadays the one cosmological scheme that is almost universally accepted among astronomers and physicists. The Big-Bang theory's success in this is a result of its predictive nature. Just like relativity, the theory of the expansion contains the seeds of observational predictions, which have all, more or less, been confirmed subsequently, following advances in observational instrumentation.

We can, in our imagination, run the film of the expansion of the universe backwards, and watch galaxies move backwards in time. Even better, by assuming that the velocity of expansion has remained constant over time, all we need do is take the reciprocal of the Hubble constant to obtain the age of the universe! If the Hubble Constant is around

■ THIS IMAGE OBTAINED BY THE HUBBLE SPACE TELESCOPE IS EXCEPTIONAL. IT SHOWS TO PERFECTION THE STARS OF MAYALL 2, A GLOBULAR CLUSTER BELONGING TO THE ANDROMEDA GALAXY, M 31, AT 2.5 MILLION LIGHT-YEARS. THE TWO BRIGHT STARS THAT SEEM TO FRAME IT, LIE IN THE MILKY WAY.

■ NGC 604 IS THE LARGEST
KNOWN NEBULA IN THE LOCAL
GROUP. IT GLOWS AT THE END OF
ONE OF THE SPIRAL ARMS OF M 33
IN TRIANGULUM, AT A DISTANCE OF
2.7 MILLION LIGHT-YEARS. IT IS
1000 TIMES AS BRIGHT AS THE
FAMOUS ORION NEBULA, M 42.
TWO HUNDRED YOUNG STARS,
15 TO 60 TIMES THE MASS OF
THE SUN, ILLUMINATE THE
NEBULA, HERE
PHOTOGRAPHED BY THE
HUBBLE SPACE
TELESCOPE.

THIS PHOTOGRAPH SHOWS PART
OF THE DISK OF THE ANDROMEDA
GALAXY, M 31. THIS BEAUTIFUL
SPIRAL IS POPULATED BY HUNDREDS
OF BILLIONS OF STARS SIMILAR TO
THOSE IN THE MILKY WAY. ONLY
A FEW HUNDRED OF THEM, BLUE
AND RED SUPERGIANTS, ARE
VISIBLE HERE. BUT EACH SQUARE
MILLIMETRE OF THE GREY MIST
THAT FORMS THE DISK OF
M 31 IN THIS PICTURE
ACTUALLY CONSISTS OF
NEARLY A MILLION
STARS...

75 km/s/Mpc, this 'Hubble Time' amounts to around 12 billion years. Naturally, this is just a very rough estimate of the time that the universe has existed. We shall see later that, for one thing, the exact value of the Hubble constant is still unknown; for another, that it probably varied over the course of time; and finally, that other cosmological factors need to be taken into account in the equations that allow us to retrace the evolution of the universe. A number of physicists, as we shall see later, prefer to set the age of the universe at about 15 billion years. For all that, this figure of 12 or 15 billion years is extraordinary on more than one count. For a start, it seem incredibly low. Judge for yourself: the study of stellar evolution has shown that the least massive stars, the red dwarfs, thanks to the rate at which they are consuming their energy reserves, have a life-expectancy of several tens of billions of years. Do we know any stars that are that old? No. In our galaxy, the oldest objects with accurately known ages are the globular clusters that orbit in the halo. Their hundreds of thousands of stars were all born at the same time from a single concentration of gas, and have each evolved as a function of their initial mass. During the course of their life, there is a very specific moment when they cease to convert their central hydrogen into helium and change their nuclear regime, becoming red giants. Physicists know, to a high degree of accuracy, when a star of a given mass becomes a red giant. A very massive star, such as Rigel, Deneb, Betelgeuse, or Antares, reaches this stage after just a few million years. A moderately massive star, such as the Sun, becomes a red giant after 10 billion years, a star of 0.8 solar mass in 20 billion years, one of 0.1 solar mass in 50 billion years, etc. In a globular cluster, thanks to the statistical study of several thousand stars, it is possible to determine the mass of the stars that are just about to become red giants, and this gives a good estimate of the cluster's age. In the Milky Way, the ages of globular clusters lie between 10 and 15 billion years, which is a figure astonishingly close to the Hubble Time, as estimated from the universe's speed of expansion.

■ For several decades astronomers have been discovering that the universe has a profound unity. This photograph shows the disk of the galaxy M 33 in Triangulum, a spiral lying near M 31 in Andromeda. The stars in this galaxy are identical to those in the Milky Way.

There is another method of dating stars, which consists of measuring the brightness of white dwarfs. Once an old star, after having passed the red-giant stage, has shed the gaseous atmosphere that surrounds it, and thus lost most of its mass, all that remains is its tiny, bright core. The former red giant has become a white dwarf. Such a dying star has too low a mass, and is too rich in the heavy elements that it has created during its lifetime, for it to be capable of sustaining any nuclear reactions. Its surface temperature, and thus its luminosity, slowly decline over the course of time, like a dying ember on the hearth. If stars had existed for all eternity, then we would find white dwarfs at every stage of cooling. Yet astronomers have never found a single white dwarf in the Galaxy that is older than some ten billion years. If we add to this age the time that the star spent on the Main Sequence, we again arrive at some 15 billion years as the age of the oldest stars known.

THE PRIMORDIAL NUCLEOSYNTHESIS

The expansion of the universe, as measured by the redshift of galaxies, is the principal evidence in favour of the Big-Bang theory, but the latter is also based on two other observational facts that are of crucial importance: the primordial nucleosynthesis of light elements, and the cosmic background radiation.

Stars are nuclear reactors, which continuously burn the hydrogen of which they are largely composed, turning it into helium, and then, in the most massive stars, as their internal pressure and temperature increase, into heavier and heavier elements. The first such elements are oxygen and carbon, then sodium, neon, magnesium, sulphur, calcium, silver, nickel, and silicon up to iron, the most stable element in the universe. Physicists, after having measured the spectra of stars, of the shells ejected by supernovae, and of the nebulae where new stars are born, nowadays know the overall composition of the universe. It consists, in round figures, of 75% hydrogen, 24% helium, and 1% for all the other elements. The second piece of evidence supporting the Big Bang is these three figures. Hydrogen nuclei are considered by all astronomers to be primordial. Consisting of a single proton, hydrogen is the simplest of all the elements; it would have been created in the flood of energy released by the Big Bang. But its existence is not a proof of the validity of the theory, because it is possible to imagine numerous other cosmological models in which the universe would also consist primarily of hydrogen. The strength of the Big-Bang theory lies elsewhere: only it is able to account for the enormous difference between the abundance of helium and all the other heavier elements. The very low percentage of the latter is readily explained if the universe is relatively young, as the Big-Bang theory suggests. But what about helium? If this element had been produced solely by the fusion of hydrogen in stellar

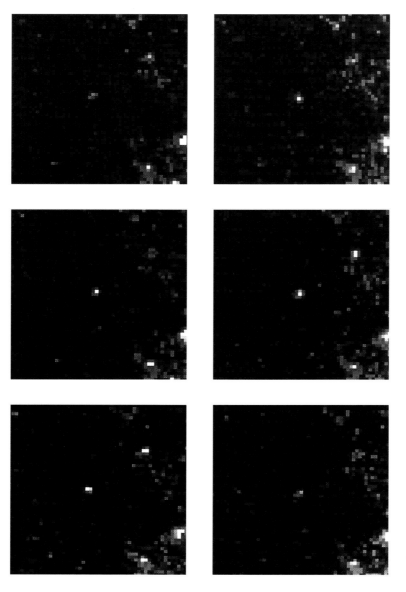

■ ASTRONOMERS ARE ABLE TO FOLLOW THE SLOW PULSATION OF A CEPHEID IN THE GALAXY M 100 (SEE PHOTOGRAPHS ON P.90 AND 95). ON THESE IMAGES, TAKEN AT 14-DAY INTERVALS, THE CYCLIC VARIATIONS OF THE STAR ARE EASILY VISIBLE. CEPHEIDS ALLOW US TO DETERMINE THE DISTANCE OF GALAXIES, AND TO ESTIMATE THE HUBBLE CONSTANT, WHICH HAS STILL NOT BEEN DETERMINED ACCURATELY.

the Big Bang, the temperature of the nascent universe is around 1000 billion degrees, but one second later it is no more than 10 billion degrees. After three minutes, the universe begins to behave like the core of a hot star: around 1 billion degrees, protons and neutrons combine to produce the first nuclei of atoms – hydrogen and helium. Scarcely two minutes later still, this primordial nucleosynthesis ceases abruptly. The cosmological expansion has lowered the temperature to a few hundred million degrees, literally freezing the hydrogen and helium that had been created. To within a few per cent – thanks to more than 10 billion years of stellar nucleosynthesis – their abundances remain precisely the same today.

THE COSMIC BACKGROUND RADIATION

The universe, in that distant past, was a form of

interiors at the same rate as they produce heavier elements, there would be just a few per cent! Why is helium so abundant in the universe? To try to understand the answer, physicists have attempted to retrace the history of the universe, from the initial singularity onwards, by drawing inferences about the behaviour of the radiation and matter present within it, and considered as a fluid, whose pressure and temperature were continuously decreasing. The precision – which at first sight appears almost surreal – with which they are able to describe events that occurred more than ten billion years ago, in a medium that was denser and hotter than the heart of a giant star – a medium that it is impossible to recreate even in particle accelerators – may seem somewhat surprising. To astrophysicists, however, the primordial universe is no more complex that the nuclear core of stars – and stellar nucleosynthesis is probably one of the best-understood of all natural phenomena. Unlike the searing material within stars, which stays in the same physical state for millions or even billions of years, the flood of energy that emerges from the Big Bang is subject to a continuous drop in temperature as space expands. One millionth of a second after

cosmic soup, an inextricable mixture of energy, radiation, and particles, a sort of searingly hot and opaque plasma, like the centre of the Sun. Particles were flying in all directions, colliding at velocities close to the speed of light, creating here a nucleus of helium (two protons and two neutrons), and there a nucleus of deuterium (one proton and one neutron), nuclei that, because of the high temperature, were unable to capture a passing free electron to create a true atom.

As the expansion continued, the density and the temperature of the universal soup also continued to decline. The age of the universe began to be counted in days, then years, in centuries, and millenia. When the great cosmic clock reached 300 000 years, the universe changed into another state. The temperature, which had declined to 3000 degrees, equivalent to the surface of a cool star, was sufficiently low to allow atoms to form, and nuclei began to capture free electrons. It was like a dense fog of electrons that slowly began to lift, allowing the intense radiation emitted by the Big Bang to travel freely throughout the expanding space. From being an opaque plasma, the universe reached the stage of being a transparent gas, freeing a veritable sea of light. In 1948, the American physicists George Gamow, Ralph Alpher, Robert Herman and Robert Dicke predicted the existence of this cosmological radiation, which is inevitably present in any theory of a universe that expands from an extremely condensed state. If any witness could have existed at that distant epoch, the universe would have appeared as a sea of light, with an intensity and colour similar to that of our own Sun, but a Sun whose radiation came from all directions, uniformly covering what would much later become the celestial sphere.

The cosmic background radiation always exists. Emitted in all directions at the moment the universe became transparent, it fills all of space, and bathes us permanently in its fossil light. But its properties have changed. In the fifteen-odd billion years that have passed, expansion has increased the scale of the universe by a factor of 1000. According to relativity, the dilation of space is accompanied by an increase in the wavelength of radiation crossing it. Let us note in passing that it is by this relativistic effect that astronomers explain the redshift: the increase in the wavelength is proportional to the length of time that the light has travelled through the universe. Nowadays, the intense light emitted by the Big Bang no longer appears in the form of light, but as microwave radiation. Emitted at a wavelength of about 800 nanometres, it now has a wavelength of about 0.8 mm!

The cosmic background radiation was discovered accidentally in 1965 by Arno Penzias and Richard Wilson, two young American researchers who were testing an aerial designed to pick up the signals from the first artificial satellites. To Penzias and Wilson, it appeared as 'radio noise' arriving from every part of the sky, and the two researchers initially thought that it was some from of parasitic noise. However, they went on to make one of the most important discoveries in the history of cosmology, bringing spectacular confirmation, together with the expansion and the nucleosynthesis of light elements, of the theory of the Big Bang.

■ THIS SUPERNOVA, WHICH IS SLOWLY FADING NOT FAR FROM THE NUCLEUS OF ITS PARENT GALAXY, IS ONE OF THE MOST DISTANT STARS EVER OBSERVED. ITS DISTANCE IS AROUND 3 BILLION LIGHT-YEARS. EXTRAGALACTIC SUPERNOVAE, LIKE THE CEPHEIDS, ARE USED BY ASTRONOMERS AS STANDARDS TO MEASURE THE HUBBLE CONSTANT.

For cosmologists, the entire history of the universe is summed up in the expansion of space, which, from the initial singularity, leads to a decrease in the pressure and temperature of the universe, which may initially be considered as an energy flux, then a searingly hot plasma, and finally as a gas that becomes less and less dense. Today, the universe is essentially empty, its matter having condensed into galaxies. If the universe were completely empty, its temperature would be that of absolute zero, 0 Kelvin, or −273.16 °C. The fossil radiation discovered by Penzias and Wilson, which earned them the Nobel Prize for Physics in 1978, is witness to the universe's dense, hot past, and is also a sign of its current average temperature: −270.42°C, or 2.726 K. (For simplicity, the specialists talk of the 3-K radiation.) The temperature of the cosmic background radiation, its spectral distribution, its homogeneity and isotropy are precisely those that correspond to the Big-Bang theory.

Are there, however, radically different hypotheses that would allow us, for example, to reintroduce the perfect cosmological principle that was so dear to Albert Einstein? A few scientists are still searching in various other directions. Among them are eminent physicists and astronomers such as Fred Hoyle, Jayant Narlikar, Halton Arp, Geoffrey Burbidge, and Jean-Claude Pecker, who have been fighting for decades against the Big-Bang theory, which they view as a creationist theory with strong religious overtones. Originally suggested by a cleric – Georges Lemaître was the priest in his town of Louvain – the theory has often been compared with the *Fiat lux!* of the book of Genesis. To its detractors, however, the Big Bang does not suffer just because of its simplistic resemblance to the Creation. To them, the hypothesis seems to be ad hoc, and they wonder about the physical sense of the initial singularity, which was an event without a cause, and accepted as such by scientists even though it lies – today, just as it did in 1930 – outside their field of research. Finally, these researchers – for reasons that are simultaneously scientific and philosophical – have chosen a universe that complies with the perfect cosmological principle: it had no moment of creation, and is infinite in both space and time. The most famous of these alternative models of the universe is the Steady State theory, proposed in Cambridge in 1950 by Fred Hoyle, Hermann Bondi, and Thomas Gold. In this universe, which, like the Big-Bang theory, is expanding, the galaxies are receding from one another in agreement with general relativity and the Hubble law. According to the Cambridge School, however, the expansion has lasted for all eternity and occurs in an infinite universe. In this universe, which Fred Hoyle and Jayant Narlikar still defend today, particles appear spontaneously, created by the energy of the expanding space-time field. This model, which has great formal beauty, had a degree of success amongst cosmologists during the 1950s and 1960s. This attraction was explained all the more, because at that time the Big-Bang theory was confronted with a major difficulty: measurements of the Hubble constant, which had been greatly overestimated by Hubble and his successors, led to a Hubble time, and thus an age for the universe, that was less than the age of the Earth! Subsequently, following measurements made by Allan Sandage and Gérard de Vaucouleurs, this famous constant, which was estimated by Hubble himself to be 500 km/s/Mpc, soon settled down to being between 50 and 100 km/s/Mpc, values that were compatible with the ages of the oldest stars. Finally, in 1965, the discovery of the cosmic background radiation, and the work of astrophysicists on the primordial nucleosynthesis of helium, were a fatal blow to the Steady-State Theory, which was unable to explain them. The expansion, the 3-K radiation, and the 24% of helium present in the universe are the three incontrovertible observational facts that need to be taken into account by cosmologists. For the majority of them, they are three proofs that we are living in an expanding universe that had a hot, dense origin.

THE LOOK-BACK TIME AND THE REDSHIFT

Researchers do not rely solely on the numerical results of Einstein's equa-

tions to sustain the Big-Bang theory, but also – and primarily – have recourse to astronomical observations. Light, for cosmologists, amounts to a time machine capable of travelling back in time. Because the velocity of light is finite, we see objects earlier in time, the farther away they are. When distances are merely astronomical, rather than cosmological, it is easy to map the universe using the velocity of light as a standard unit. The distance of the Moon is about 380 000 km. At a velocity of 300 000 km/s, light crosses this distance in about one second. To astronomers, the distance of the Moon is therefore about one light-second. When it comes to the stars, they are too far for their distances to be measured in kilometres. In the Milky Way, distances are reckoned in tens, hundreds, or thousands... of light-years. The difference in time that these distances imply has minimal significance from an astronomical point of view. The fact that Deneb, at a distance of 3000 light-years, is seen as it was 3000 years ago, is perfectly irrelevant for astronomers. The evolutionary lifetimes of stars are measured in millions of years, so astronomers are used to ignoring the time element introduced by the distances of stars – at least when they are nearby. They were able, for example, to follow the explosion of the supernova that appeared in the Large Magellanic Cloud in 1987 'as if' it was occurring immediately in front of their eyes. In fact, the star Sanduleak −69°202, lies at a distance of more than 170 000 light-years, a distance which corresponds to a look-back time of 170 000 years.

At greater distances – let us say, to fix our ideas, out to five billion light-years, i.e. within a sphere that corresponds, to astronomers, to the local universe – we can still continue to use the light-year as a simple, and beautiful, unit, and even, if we wish, convert it into kilometres. A billion light-years amounts, after all, to no more than 10 000 000 000 000 000 000 000 km. But farther away, talking of the distances of bodies no longer has any sense. At such cosmological scales, the expanding geometry of the universe needs to be taken into account, and forces us to abandon our notions of spatial distance in favour of one of distance in time. The

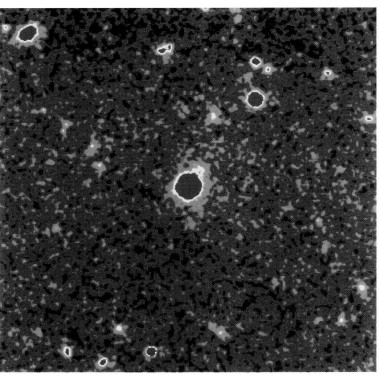

■ AT VERY GREAT DISTANCES, BECAUSE OF LIGHT'S LONG TRAVEL TIME, WHICH IS ACCOMPANIED BY A CHANGE IN THE SCALE OF THE UNIVERSE, CAUSED BY THE EXPANSION, IT IS IMPOSSIBLE TO USE THE CLASSICAL UNITS OF MEASUREMENT, SUCH AS A LIGHT-YEAR. ASTRONOMERS THEREFORE SUBSTITUTE THE REDSHIFT z. HERE, ONE OF THE MOST DISTANT GALAXIES KNOWN, THE BRILLIANT QSO 1207−07, IN THE CENTRE, HAS A REDSHIFT OF 4.4. THAT CORRESPONDS TO A LOOK-BACK TIME OF 90%.

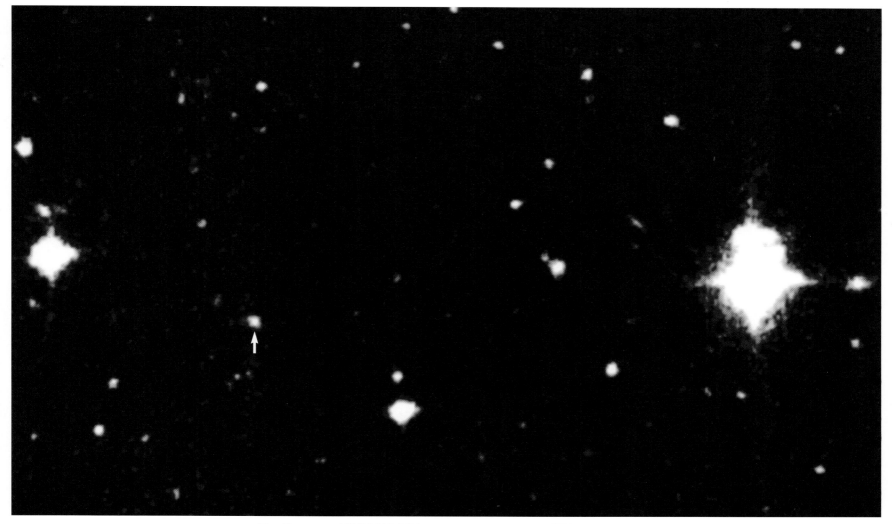

galaxy MRC 0316–257, in the constellation of Fornax, is receding at an apparent

■ THE GALAXY MRC 0316–257, OBSERVED WITH THE 3.6-M CFH TELESCOPE AT THE OBSERVATORY ON HAWAII. THE LIGHT REACHING US TODAY FROM THIS DISTANT GALAXY WAS EMITTED WHEN THE UNIVERSE WAS 15% OF ITS PRESENT AGE, THAT IS, SOME 12 BILLION YEARS AGO.

velocity close to 300 000 km/s: using the Hubble law, astronomers give it a distance of more than 12 billion light-years. This is a convenient, but inappropriate, simplification, because in the geometrical framework dictated by relativity, this figure has no real meaning. Imagine a ray of light that was emitted by this galaxy, 12 billion years ago. During the time that this photon has been propagating towards us, the universe has continued to expand, and the distance between the two bodies has increased. It is precisely the dilation of the geometry of space that has caused the redshift in the light detected by our telescopes today. What does the image of the galaxy MRC 0316–257, as photographed by astronomers, actually represent? Are we seeing the galaxy as it exists now? No – observing far out into space is observing far back in time. We see the galaxy as it was 12 billion years ago.

This distance in time is the only measurement that we have any right to use. When MRC 0316–257 emitted the photons that we are detecting today, the scale of the universe was smaller, and the galaxy was much closer. It was, at that far distant epoch, only 4 billion light-years away from the Milky Way. Is this the 'distance' that we ought to use? Or its current 'distance'? Relativity's equations indicate that MRC 0316–257 currently lies at a distance of 16 billion light-years. Yet any idea of contact between MRC 0316–257 and us today has no sense in this context. Our only link with this galaxy is with its past. In fact, when astronomers consider objects at cosmological distances, they no longer use their local methods of distance measurement (the kilometre, the light-year, or the megaparsec). Instead they substitute two new measurements:

the 'look-back time', and the spectral shift (the famous redshift). The redshift z is an absolute measurement of the change in scale of the universe, independent of the uncertainties concerning the value of the Hubble constant and the age of the universe. For the galaxy MRC 0316–257, $z = 3.14$. This figure means that at the time the light that we detect today was emitted, some twelve billion years ago, the scale of the universe $(1 + z)$ was 4.14 times smaller. Any volume within the universe was, in accordance with the Big-Bang theory, simultaneously smaller and hotter than today. The redshift z, is cosmologists' standard of measurement, because, once again, it is independent of the actual cosmological model employed. It is, however, possible to translate z into a more significant unit of measurement that is more accessible to the non-specialist: the look-back time (see the Appendices). In the case of MRC 0316–257, this look-back time is about 85%. To put it in plain English, when we look at a photograph of MRC 0316–257, we are looking back over 85% of the time that has elapsed since the Big Bang, and we are seeing the distant galaxy as it was when the universe's age was 15% of its current age. Although the equation seems more complicated and less poetic than the classic measurement of space-time distances in light-years, it is nevertheless necessary to gain a true impression of the scale of cosmological time. Astrophysicists do not currently know the actual numerical values of any of the four fundamental cosmological parameters – the Hubble constant is just the most famous – that describe the relativistic geometry of the universe, and which would enable them to describe its history completely, from the Big Bang to the present day. In fact, as we have seen, maintaining that the galaxy MRC 0316–257 is at a distance of 12 billion

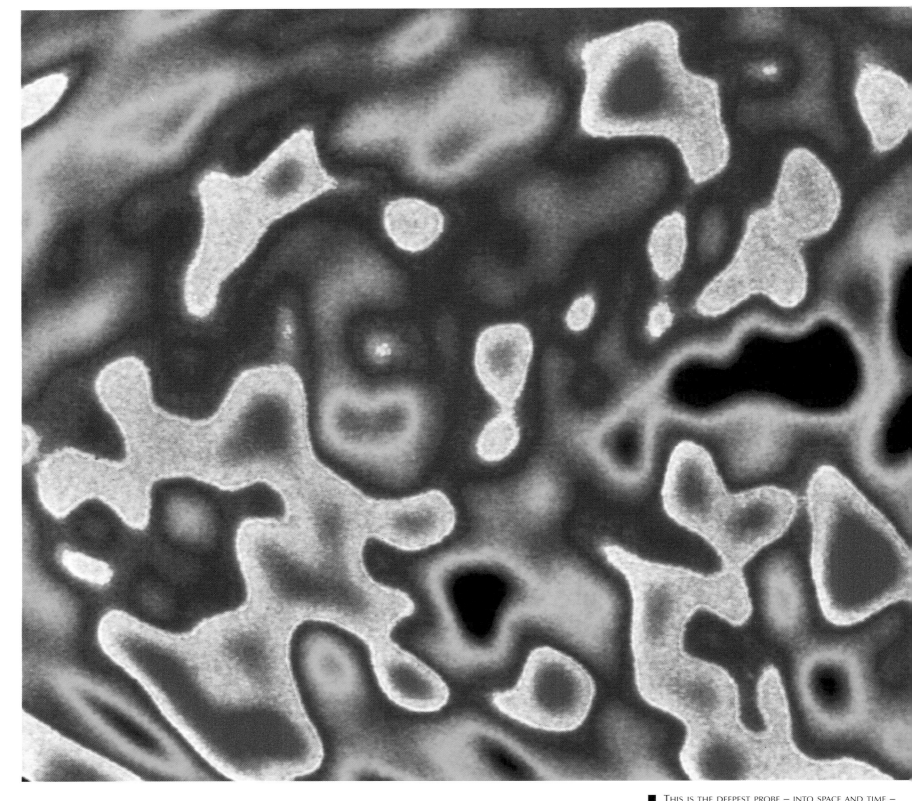

light-years has no sense, but even worse would be to say that we see this galaxy as it was 12 billion years ago, with is also an incorrect simplification. Such a period of time is valid for one theoretical model of the universe. In some other cosmological model, we might assign, just as arbitrarily, a period of 8, 10, 12, 14, or 16 billion years to MRC 0316–257. The use of the look-back time thus allows researchers to hide their ignorance of the age of the universe, by describing the objects that they are studying in terms of a simple percentage of the (unknown) overall time. While they wait for their cosmological theories to be refined, astronomers console themselves by mapping the universe, using the redshift z. For theorists, its use is reassuring: as we have seen, its value is absolute, independent of any model, and z accurately describes the change in the scale of the universe through the overall expansion.

Like metric units for measuring length, this new relativistic scale for the geometry of the universe is open-ended. At $z = 0$, we are within the local universe, where the expansion is imperceptible. The Virgo Cluster lies at $z = 0.004$, the Coma Cluster at $z = 0.03$, the cluster Abell 370 at $z = 0.37$, and the galaxy MRC 0316–257 at $z = 3.14$. The most distant galaxies so far determined are beyond $z = 5$. The greatest redshift known is that of the cosmic background radiation, $z = 1000$.

Astronomers have only been able to reach such cosmological distances quite recently, since giant telescopes and their extremely sensitive electronic cameras have been put into service. Until the beginning of the 1960s, their investigations were limited to our relatively close neighbourhood. Excursions beyond one billion light-years were very rare. Observations of the millions of galaxies that are contained within this apparent sphere, centred on the Earth, showed a universe that was homogeneous and isotropic, populated by clusters of galaxies uniformly spread throughout the darkness of space. It was

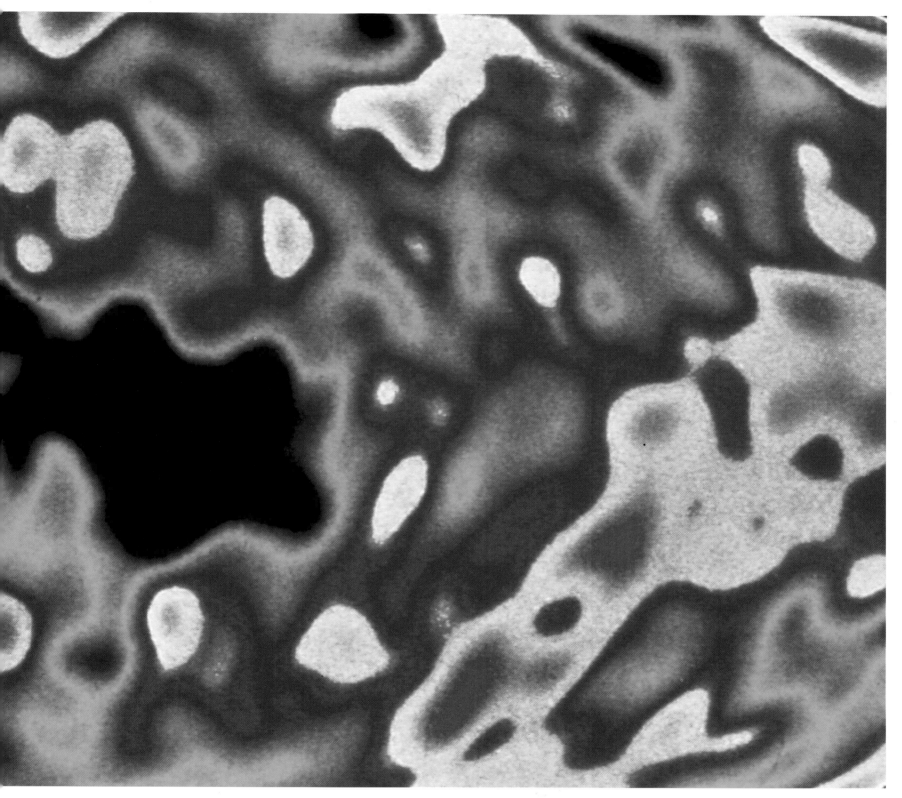

UNIVERSE JUST 300 000 YEARS AFTER THE BIG BANG. THE INHOMOGENEITIES DISCOVERED IN THIS FOSSIL LIGHT, EASILY VISIBLE IN THIS IMAGE TAKEN IN 1992, MAY PERHAPS INDICATE FLUCTUATIONS IN THE PRIMORDIAL MATERIAL WHICH WOULD LATER GIVE RISE TO GALAXIES.

extremely difficult, when contemplating these myriad galaxies floating in seemingly infinite space, not to subscribe to the infinite, eternal universe of Hoyle, Bondi and Gold. The last three decades have radically changed the perspective of the universe held by most astronomers.

The major telescopes finally revealed the universe's distant past, regions of space that lie beyond $z = 1$, which corresponds to a look-back time of around 60%, i.e. about 10 billion years.

In these ancient regions, astronomers have found galaxies that show differences in their morphology, dynamical behaviour, and stellar composition from galaxies in the local, modern-day universe. This is an observational fact of profound significance, which they have suspected for a long time, but which they were unable to detect. Like a new proof of the validity of the Big-Bang theory, they have discovered with their own eyes that the universe is evolving, that it has a history. ■

Gravitational lenses

■ GRAVITATIONAL LENSES, PREDICTED BY THEORISTS FROM GENERAL
RELATIVITY FOR MORE THAN 60 YEARS, HAD NEVER BEEN SEEN BY
ASTRONOMERS FOR LACK OF SUFFICIENTLY POWERFUL TELESCOPES. FOR MORE
THAN A DECADE, DOZENS OF THESE STRANGE OPTICAL ILLUSIONS HAVE NOW
BEEN DISCOVERED IN THE DEPTHS OF SPACE. HERE, THE CLUSTER ABELL
2390, PHOTOGRAPHED BY THE HUBBLE SPACE TELESCOPE, SHOWS
LUMINOUS ARCS ELONGATED VERTICALLY. THESE ARE THE GHOSTLY
IMAGES OF DISTANT GALAXIES.

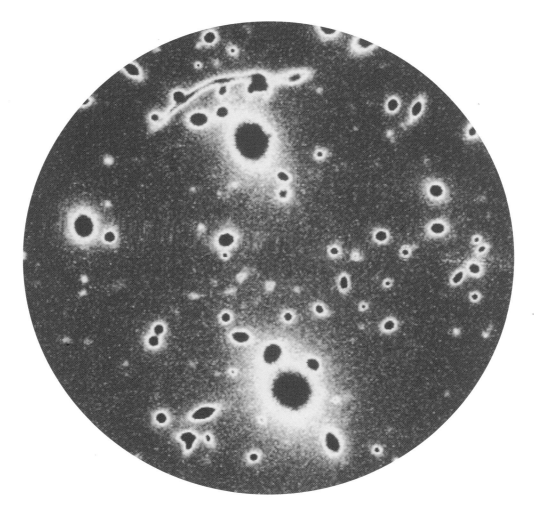

■ THE FIRST GIANT ARC, WHICH LIES ON THE BORDERS OF ABELL 370, WAS DISCOVERED IN 1985 BY FRENCH OBSERVERS FROM THE TOULOUSE OBSERVATORY USING THE 3.6-M TELESCOPE ON TOP OF MAUNA KEA VOLCANO IN HAWAII.

t is a minute area of sky, to all appearances devoid of stars, lying in the eastern part of the constellation of Cetus, on the borders of the long, sinuous constellation of Eridanus. Photographed with a large telescope – after several hours exposure with an electronic camera – this area of the sky, which appears utterly dark to the naked eye, reveals a host of small, fuzzy spots crowded together. This is the cluster Abell 370 (A 370): the most distant of the nearly 3000 clusters catalogued by the American astronomer George Abell at the end of the 1950s. This incredible extragalactic grouping, 500 000 light-years across, contains more than 100 000 billion stars, gathered into at least a hundred galaxies. As seen from Earth, however, these galaxies appear as minute fuzzy spots. To make them clearly visible, astronomers need to apply complex image-processing, before displaying them on their VDUs. The dark blue background sky is lightened, and – to amplify their contrast and luminosity – the galaxies in the cluster are converted into false colours: a range of red and yellow tints. Nevertheless, the smallest of the galaxies are lost in the sky background, and the brightest are no more than magnitude 20. The redshift, $z = 0.37$, of A 370 is very large. It corresponds, in

standard cosmological models, to a distance of around four billion light-years. At such a great distance in space-time, it is hardly surprising that these galaxies appear so small and faint, even when imaged with giant telescopes.

OPTICAL ILLUSIONS IN THE SKY

It is, however, not its distance, size or mass that has rescued A 370 from anonymity, and caused it to become one of the stars of astrophysics at the end of this millennium. Nowadays no one even attempts to count clusters of galaxies that have similar, or even more striking, physical characteristics. When, in the 1970s, astronomers began to take photographs of A 370 with large telescopes, the images of the cluster revealed a curious curved structure towards one edge that was very elongated and, above all, extremely faint. So faint, in fact, that it escaped most observers. In particular, its appearance was so strange that those who did notice it decided that it was not worth studying, easily convinced that this pale streak was nothing more than a reflection on the photographic plates or an artefact produced by the recently introduced, and still recalcitrant, electronic cameras. In fact, in 1973, 1977, 1981, and finally even in 1982,

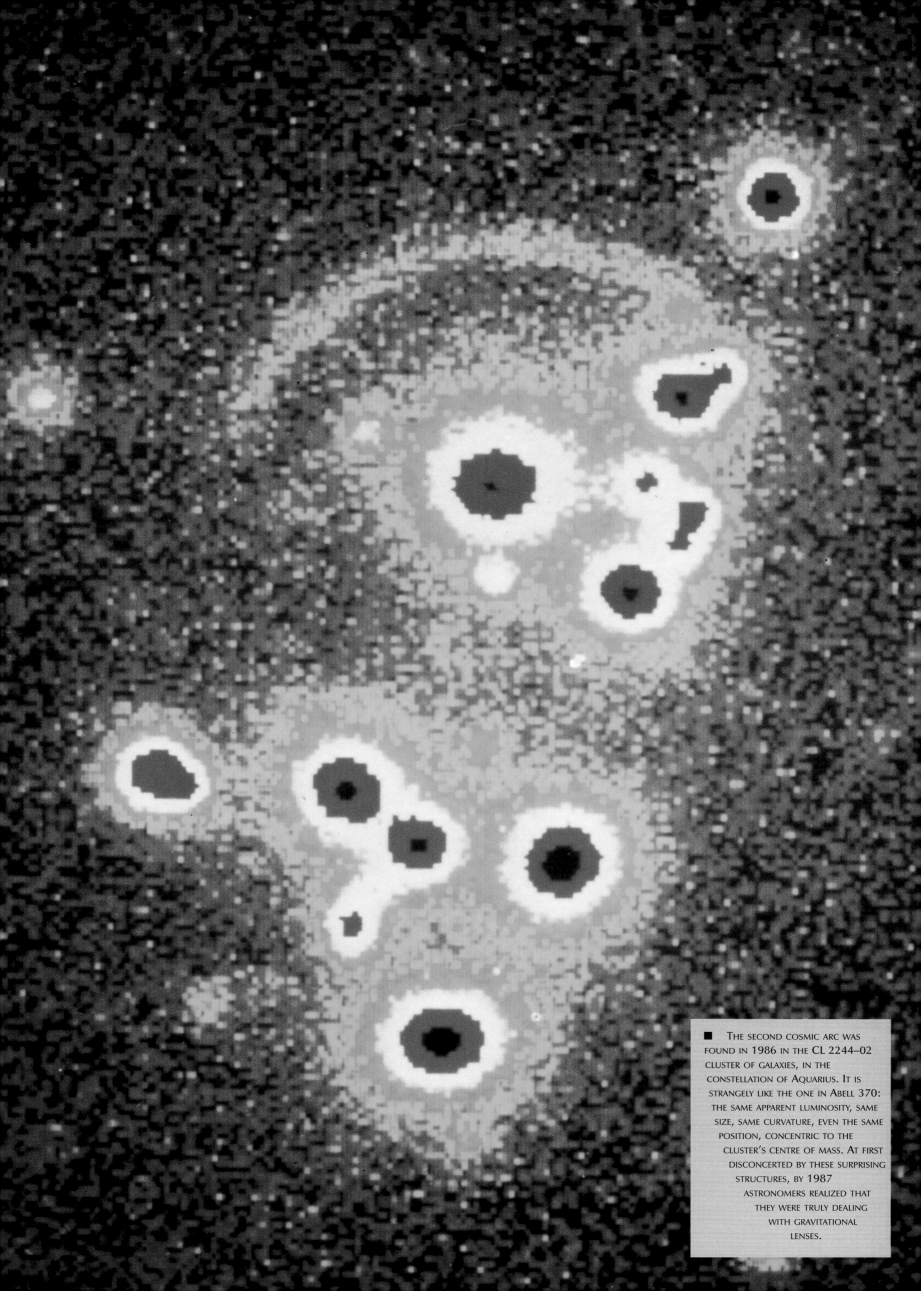

■ THE SECOND COSMIC ARC WAS
FOUND IN 1986 IN THE CL 2244–02
CLUSTER OF GALAXIES, IN THE
CONSTELLATION OF AQUARIUS. IT IS
STRANGELY LIKE THE ONE IN ABELL 370:
THE SAME APPARENT LUMINOSITY, SAME
SIZE, SAME CURVATURE, EVEN THE SAME
POSITION, CONCENTRIC TO THE
CLUSTER'S CENTRE OF MASS. AT FIRST
DISCONCERTED BY THESE SURPRISING
STRUCTURES, BY 1987
ASTRONOMERS REALIZED THAT
THEY WERE TRULY DEALING
WITH GRAVITATIONAL
LENSES.

this large structure was no sooner recorded than it was forgotten.

In September 1985, when the 3.6-m diameter Canada-France-Hawaii Telescope, sited at the top of Mauna Kea volcano in Hawaii, was itself turned towards the constellation of Cetus, it was fitted with a CCD camera, whose sensitivity far exceeded photographic plates or the first electronic detectors. The long blue streak surrounding A 370 that the French astronomer Bernard Fort rediscovered was so readily visible that it became impossible to ignore it any longer. But what was it? Several unconvincing ideas were initially put forward: the arc could have been produced by a colossal explosion that took place in a giant galaxy in the centre of the cluster. But what sort of body, what sort of event, could account for the expulsion of a shell of plasma more than 100 000 light-years in radius? There was also the even stranger possibility that Abell 370's blue arc was a light-echo, i.e. the reflection of the flash from a very ancient supernova explosion, propagating in a diffuse medium that surrounded the cluster. Such a phenomenon was, in fact, discovered three years later, after the explosion of the supernova in

the Large Magellanic Cloud (see Chapter 7), but the idea could not be sustained for A 370. At the cluster's distance of 4 billion light-years, no supernova explosion could have caused such a powerful flash of light. The solution emerged suddenly, more than two years later, when the American astronomers discovered another bluish luminous arc, almost identical to the one in A 370, in a cluster of galaxies in Aquarius, Cl 2244–02. In both cases, the luminous arc appears strangely symmetrically placed in the cluster. It is exactly centred on the core of the latter, where the most numerous and most massive galaxies are located. The theorists realised that these two arcs were not some new sort of object, as large as galaxies and somehow ejected from these distant clusters. In addition, there is no physical process in the universe capable of releasing so much energy and creating such structures. In fact, the bluish arcs do not really exist, they are simply optical illusions... By chance, the astronomers had come across one of Albert Einstein's most amazing predictions, half-expected by theorists for more than fifty years: a gravitational lens.

According to Einstein's theory of general relativity, space-time is curved by the mass that it contains. General relativity therefore predicts that when passing close to a very massive object – a star, for example – the path of a ray of light becomes curved, following what is known as a geodesic. The ray of light is thus deviated by the star's gravitational field. As seen from the Earth, projected against the celestial sphere, this curved path is marked by an infinitely small change in the position of the object against the background sky. When the theory of general relativity was published in 1915, no one knew anything at all about the mass, diameter, or distances of the galaxies, or about the large-scale structure of the universe. The only possible proof of this strange prediction of Einstein's had to be sought among celestial objects that were well-understood by astrophysicists. So the stars served as the distant reference points, and the Sun, whose mass and distance are known with great accuracy, served as the perturbing body. In its apparent motion across the sky, the Sun appears to pass in front of the stars in the twelve constellations of the zodiac. If the general theory of relativity is valid, the rays of light from those stars should, before they reach the Earth, be shifted slightly by the gravitational field of our star. Einstein even calculated the theoretical value of this deviation at the edge of the solar disk. The tiny angle corresponds to moving the tip of a pen, seen at a distance of 100 m, by just 1 mm. But how could such an experiment be carried out? How could you measure the positions of stars accurately in full sunlight? Astronomers devised an

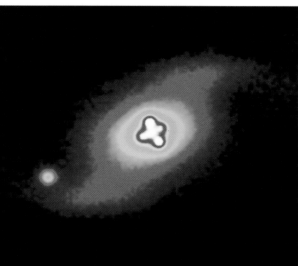

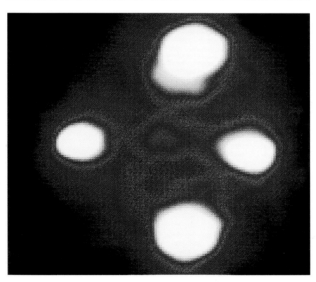

■ THE EINSTEIN CROSS IS PROBABLY THE MOST SPECTACULAR OF THE GRAVITATIONAL IMAGES. IN THIS CASE, THE OBJECT FORMING THE GRAVITATIONAL LENS IS A BEAUTIFUL SPIRAL GALAXY IN THE CONSTELLATION OF PEGASUS, AT A DISTANCE OF 400 MILLION LIGHT-YEARS. EXACTLY AT THE CENTRE OF THE GALAXY, ASTRONOMERS HAVE DISCOVERED Q 2237+0305, A MULTIPLE QUASAR, THE FOUR COMPONENTS OF WHICH ARE ARRANGED IN A LOZENGE SHAPE AROUND THE NUCLEUS OF THE GALAXY. THIS SERIES OF IMAGES OF Q 2237+0305 WAS OBTAINED WITH THE 3.6-M CFH TELESCOPE IN HAWAII.

night. Physicists and astronomers impatiently awaited the eclipse of 29 May 1919, which was observed in this manner by Arthur Eddington, who confirmed that around the Sun all the stars were slightly displaced by an amount that – allowing for the uncertainties in measurement – exactly agreed with Einstein's prediction.

After this famous Einstein eclipse, theorists turned their attention to the strange optical effects that the curvature of space might generate on a scale larger than that of the Solar System. Various theoretical studies followed and gave rise to a new discipline, that of gravitational optics. Space, in a region of the universe that has a strong concentration of matter, in fact behaves like a lens. The analogy between the optical and gravitational domains is not purely fortuitous: the curvatures produced by media with different refractive indices, and those caused by different masses in a gravitational field, create practically identical effects.

Theorists calculated that rays of light from distant objects, perturbed by these 'gravitational lenses' might – depending on the various geometrical configurations and as a function of the apparent diameters and the curvature of the gravitational field – produce gravitational illusions. Images that were double, triple, quadruple, displaced, enlarged, or completely deformed, might be created in the sky. But did these beautiful geometrical figures imagined by the mathematicians actually exist? And if so, were these optical illusions actually observable? In fact, were astronomers not already observing these effects, without knowing it? Finally, if these images did exist, how could they tell the real objects from their ghostly doubles?

elegant and spectacular solution. All it takes is to measure the precise co-ordinates of stars close to the Sun during a total eclipse, then compare them with the co-ordinates obtained six months later, when the stars are visible in the middle of the

In fact with hindsight it would seem, from the lack of

commitment shown by observers in tracking down gravitational lenses, that most of them had little confidence in the power of their telescopes, nor did they really believe the strange predictions made by the theorists.

Finally, at the end of March 1979, almost exactly sixty years after the Einstein eclipse, an American team led by D. Walsh accidentally discovered, in the constellation of Ursa Major, the first gravitational lens, the object Q 0957+561, soon known as the 'double quasar'. In appearance, this

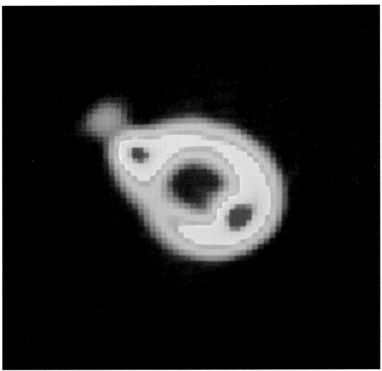

■   THE FIRST EINSTEIN RING WAS DISCOVERED IN 1987 WITH THE VLA RADIO INTERFEROMETER, WHICH IS AT AN ALTITUDE OF 2200 M, IN THE PLAINS OF ST AUGUSTIN, IN NEW MEXICO. THIS RADIO IMAGE SHOWS MG 1131+0456 AT A WAVELENGTH OF 20 CM.

looked like two point sources, identical, very faint, and very close to one another. The astronomers first thought that they were dealing with two distinct objects, fortuitously appearing close to one another in the sky by a simple effect of perspective. The sky is covered with thousands of quasars, millions of galaxies, and billions of stars, so there are innumerable apparent associations between objects that are, in reality, at vastly different distances. But the spectra from the

two objects amazed the specialists: quite apart from their very high redshift, (z = 1.4), they were absolutely identical. Just as there are no two sets of identical fingerprints, it is impossible to imagine that the complex and subtle information contained within an astronomical spectrum could be repeated. D. Walsh realised immediately that he was looking at the image of a distant quasar, doubled by an intervening gravitational lens. This was, in fact, discovered some months later. It turned out to be a galaxy, with redshift z = 0.36, lying at a distance of about 4 billion light-years. The quasar itself is at least twice as far away.

At the beginning of the 1990s, when the bluish arc in Abell 370 was independently observed at several observatories, neither the stellar displacements found by Eddington, nor the double quasar were unknown to the specialists. Yet none of them dared to suggest that these spectacular structures are (gravitational) optical illusions. It was only when a similar arc was discovered in the cluster Cl 2244–02 that the theorists took the plunge and announced that the two arcs are, in fact, the images of distant objects, radically deformed by massive, foreground gravitational lenses. There is a convincing explanation for this apparent coincidence: the clusters A 370 and Cl 2244–02 were observed because they were exceptionally rich. In addition, in both cases, the arcs appeared to be centred precisely on the dynamical centre of each cluster, that is, on their centre of mass. Finally, the specific shape of these two gravitational images could be explained if the line Earth-lens-source were slightly decentred. In a geometrically perfect case, the image of the distant object would be a perfect circle. But astronomers, delighted by their unexpected finds in two clusters of galaxies, never even dreamt that they might be able to observe this ideal shape, proof, so to speak, of the validity of relativity, and which they reverently baptised the 'Einstein Ring'. Once the calculations had been made, the mass of the gravitational lens that could produce the arc in A 370 proved to be above 10 000 billion solar masses. Finally, in 1988, the spectrum of the arc, z = 0.724, was obtained by Genevieve Soucail at the European Southern Observatory at La Silla in

■   THE MERLIN NETWORK OF AERIALS, IN GREAT BRITAIN, WHICH IS SEVERAL HUNDRED KILOMETRES ACROSS, WAS REQUIRED TO OBTAIN THIS RADIO IMAGE OF THE SECOND EINSTEIN RING, B 0218+35. THIS OBJECT IS TOO FAINT, AND TOO FAR AWAY, AT z = 0.96, TO BE SEEN BY OPTICAL TELESCOPES.

Chile, and compared with that from the galaxies in the cluster, $z = 0.375$. The measurement of the redshift as twice that of the cluster was formal proof that the source – a galaxy – lay far in the background, several billion light-years beyond the lensing cluster.

GRAVITATIONAL LENSES CONFIRM RELATIVITY

This spectrum was convincing evidence for even the most hesitant scientists that the theory of gravitational lenses was correct. They were going to have to become accustomed to the idea that Einstein's curvature of space-time was not just a theoretical, abstract model, but a tangible reality, that they would be detecting ghostly galaxies and multiple quasars, and that occasionally they would be misled by the illusions. For the last group of sceptics there remained, however, one final argument – a crucial one – against the theory of gravitational lenses. If the relativistic hypothesis was correct, then, with the constantly improving power of telescopes and bearing in mind

that astronomers had not previously been looking for them, other gravitational lenses should be discovered in the sky, particularly in fields containing the most massive clusters of galaxies. To the specialists' great surprise, effectively all the geometrical forms calculated by the gravitational-optics theorists were about to be found. After the famous double quasar Q 0957+561, astronomers discovered a second binary quasar, UM 673, in 1987. Then they became interested in a new attraction, the triple quasar, PG 1115+080, which displayed three ghostly images. Then discoveries multiplied. In 1988, a group of Belgian

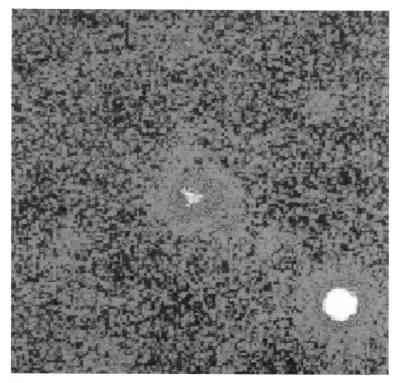

■ IN 1994, THE 10-M KECK TELESCOPE ON HAWAII MANAGED TO OBTAIN THIS INFRARED PHOTOGRAPH OF MG 1131+0456. IN THIS IMAGE, TAKEN AT A WAVELENGTH OF 1.2 MICRONS, ONLY THE FOREGROUND GALAXY, RESPONSIBLE FOR THE LENSING IS VISIBLE.

astronomers that had undertaken a systematic search for gravitational lenses, found yet another astounding object in the constellation of Auriga: H 1413+117, the quadruple quasar, soon known as the Four-leafed Clover. This quadruple quasar was the first of a class of objects that astronomers called an Einstein cross. These four images, lying in a diamond formation, all showed exactly the same spectrum, with $z = 2.55$. On the other hand, in this case, observers have yet to find any trace of the deflecting object, which is probably an extremely distant, and hence invisible, galaxy. Again in 1988, the last, and most sought-after member of the gravitational menagerie was uncovered: an Einstein ring. Theoretical models showed that if an extremely distant point source was precisely aligned with the object causing the deviation, the gravitational lens creates an image in the form of a perfect ring. The arcs in A 370 and Cl 2244–02 were only segments of rings, approximations of perfect geometrical figure, whose discovery would probably have delighted the father of relativity. The first Einstein ring, MG 1131+0456, was discovered in the constellation of Leo by the VLA radio-interferometry network in New Mexico.

And what about the arcs? They proved, in the end, to be the most numerous of all, and the easiest to detect. After those in A 370 and Cl 2244–02, arcs were found in fifty-odd clusters of galaxies! These clusters are almost all at the same distance from Earth, with redshifts z between 0.2 and 0.4. The distribution of these gravitational lenses on the surface of such a sphere centred on

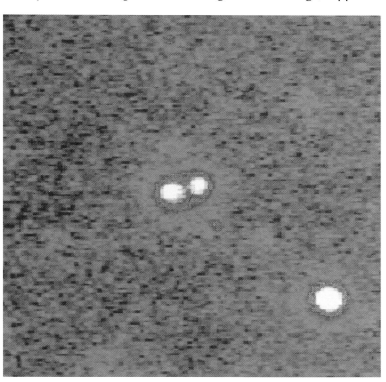

■ THIS SECOND INFRARED IMAGE OF MG 1131+0456, AGAIN TAKEN WITH THE 10-M TELESCOPE ON HAWAII, WAS MADE AT A WAVELENGTH OF 2.2 MICRONS. THE FOREGROUND GALAXY HAS BEEN SUPPRESSED TO REVEAL THE TWO BRIGHT COMPONENTS OF THE EINSTEIN CROSS.

the Earth is not accidental. Relativity theorists predicted, even before the arcs were discovered, that this was the ideal distance for the formation of images of objects that were even more remote.

It was the spectroscopic study of Cl 2244–02 that was to hold the greatest surprise for observers and gravitational theorists alike. Its distance, determined from its redshift $z = 0.33$, is around 4 billion light-years. In 1991, however, Yannick Mellier succeeded in recording the spectrum of its beautiful arc, in showing that it belonged to a galaxy, and measuring the redshift as $z = 2.23$. This value, for a galaxy, is enormous. Currently we know of only about a dozen galaxies with larger redshifts, such as the most distant ones known, 4C 41.17 and 8C 1435+63. But the latter are probably huge objects, and abnormally bright, whereas the galaxy that creates the arc in Cl 2244–02 is undoubtedly normal. Statistically, the probability that an abnormal object lies along the axis of a gravitational lens is extremely low. Yet the distance in time at which we are seeing this galaxy, in a model where the age of the universe is 15 billion years, exceeds 10 billion years! How can this galaxy be visible over such a great distance?

In fact, this observation was proof that the image of the distant source was not only deformed by the gravitational well created by Cl 2244–02, but that it had also been magnified and, above all, amplified. A final relativistic effect, predicted by the astronomer Laurent Nottale, was confirmed. The deflecting mass in the foreground that formed the gravitational lens behaved exactly like an optical lens. It concentrates the rays

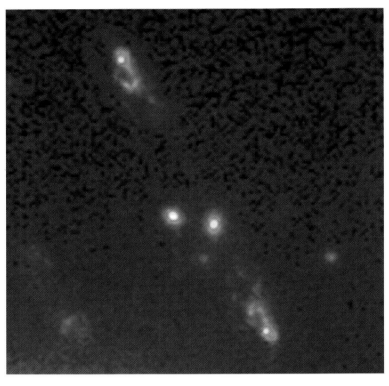

■ THIS IS THE MOST BEAUTIFUL EXAMPLE OF GRAVITATIONAL LENSING FOUND IN THE CLUSTER AC 114. THE IMAGE OF A DISTANT GALAXY APPEARS DOUBLED AND INVERTED, AS IF SEEN IN A MIRROR. THE REDSHIFT OF THE LENSED OBJECT IS $z = 1.86$, WHICH MEANS THAT WE ARE SEEING THIS GALAXY WITH A LOOK-BACK TIME OF 75%.

of light, and increases the brightness of the object that it is focussing! Although for certain geometrical configurations of the source, lens and observer the deflection of the rays of light may cause the opposite effect, i.e. one where the light is diminished, other configurations cause the light to be amplified by ten, twenty, or even one hundred times. This is obviously the case with Cl 2244–02, which provides astronomers with the disturbing image of an object that would otherwise have remained permanently invisible.

With the Hubble Space Telescope, the search for relativistic effects caused by the curvature of space took on a new dimension. More powerful and more precise, the HST enable astronomers to distinguish clearly the gravitational arcs from the host of objects – stars and galaxies – among which they had hitherto been lost as indistinct patches of light on images obtained with ground-based telescopes. This was how the field of the cluster Abell 2218 in the constellation of Draco, revealed more than 200 tiny arcs in a single HST image.

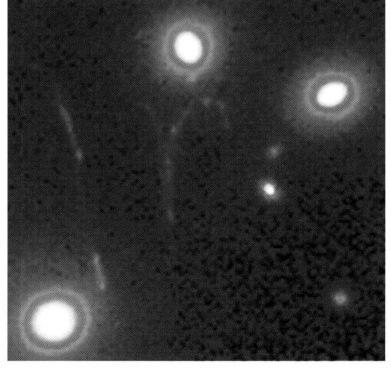

■ THE ACCURACY OF THE HUBBLE SPACE TELESCOPE'S IMAGES HAS ALLOWED US TO DISCOVER STRANGELY SHAPED GRAVITATIONAL ARCS IN THE CLUSTER AC 114. THE DISTORTIONS CAUSED BY THE CURVATURE OF SPACE DO NOT YET ALLOW US TO RECONSTITUTE THE DEFORMED IMAGES OF THESE GALAXIES, WHOSE DISTANCES ARE ALSO UNKNOWN.

PROBING THE DEPTHS OF THE UNIVERSE

These arcs are the amplified images of exceptionally distant back-ground galaxies. Although Abell 2218, for example, has a redshift $z = 0.175$, corresponding to a distance of some 2 billion light-years, the most distant galaxies that its vast mass is bringing to a focus and imaging are beyond $z = 3$, an enormous distance in space-time – equivalent to a look-back time of more than 80% – and one which telescopes are generally unable to attain. Only the magnifying effect of the foreground cluster

■ The cluster AC 114 lies in the constellation of Perseus. It was photographed in 1996 with the Hubble Space Telescope, revealing new types of gravitational lensing, here forming images with highly complex geometry, and also of amazing symmetry. The redshift of this cluster is z = 0.31; its distance is around 4 billion light-years.

enables us to see the distant universe in this manner.

The amplification of the light, and the enlargement caused by these 'gravitational telescopes' are so spectacular that at the beginning of the 1990s, astronomers started to use extremely massive clusters of galaxies to discover what the distant universe looks like. Subsequently, they found that they could distinguish between young, blue, spiral galaxies and older, redder, ellipticals, at distances where, normally, they would be invisible.

But this is not all. Thanks to the considerable degree of enlargement, images of distant galaxies show up to ten times as many details. By using several arcs – known to be several images of the same galaxy – and by modelling the shape of the foreground gravitational lens as accurately as possible, astronomers hope that they will soon be able to reconstruct detailed images of galaxies in

formation, seen as they were some 10 to 12 billion years ago. What is at stake with this research is extremely significant. In effect, although astronomers know a few galaxies that lie between $z = 3$ and $z = 5.3$, these are objects that are mostly abnormally bright, and extremely rare. Galaxies that are amplified by gravitational lenses are, by contrast, extremely numerous, and distributed completely randomly across the sky and are thus representative of the population of galaxies that existed in the distant past.

The current rate of discovery of gravitational lenses is accelerating, and the number of arcs that will be discovered in the years to come may be estimated at several thousands. In 1995, for example, the American astronomers Eric Ostrander and Richard

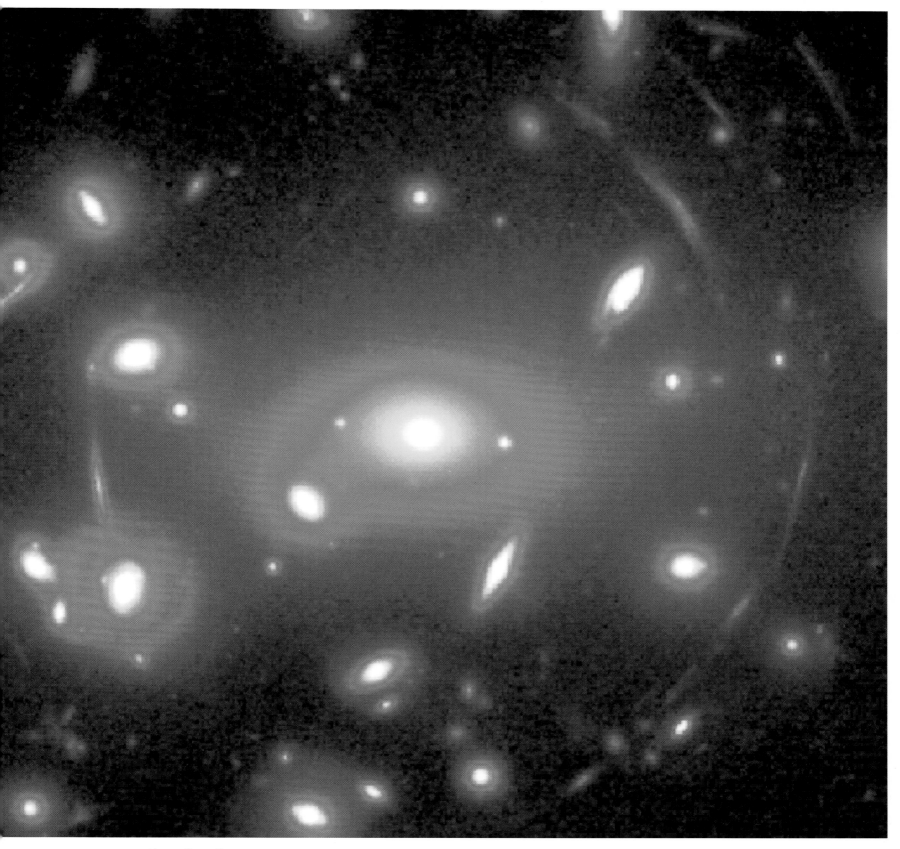

BEEN PHOTOGRAPHED BY THE HUBBLE SPACE TELESCOPE.
DOZENS OF GRAVITATIONAL ARCS SURROUND THE GIANT
ELLIPTICAL GALAXIES IN THE CLUSTER.

Griffiths discovered two new Einstein crosses on HST images. Given the small size of the field in which these two new images with a very specific shape were discovered, the researchers estimate that there may be as many as 1 million Einstein crosses and more than 10 million arcs to be detected over the whole sky.

In reality, people will probably not bother to continue cataloguing gravitational lenses after the beginning of the next millennium. Researchers begin to suspect that at great distances in the universe, all objects that are observed have been subjected to gravitational lensing. If astronomers soon cease to compare these images with one another, it will simply be because, in certain regions of the sky, they are as innumerable as all the galaxies. ■

The mystery of the missing mass

■ In the constellation of Eridanus, the rich cluster MS 0451–0305 reveals hundreds of galaxies. A few bluish spirals are visible, floating among a myriad elliptical galaxies, which appear as small orange spots. The individual movements of the galaxies enables us to calculate the overall gravitational field of the cluster, and shows that the latter probably contains 100 times as much matter as astronomers can detect from their images. This is the mystery of the missing mass.

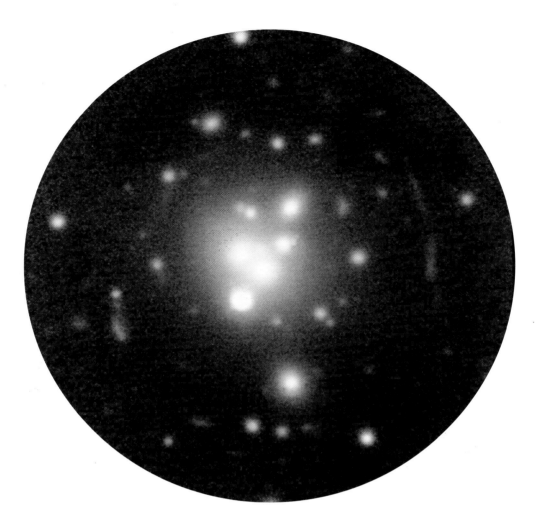

■ The cluster of galaxies MS 0440+0224, imaged with the Canada-France-Hawaii Telescope. In this cluster the ratio between the visible mass in the form of galaxies, and the mass of invisible matter is more than 1:100.

Some of the clusters of galaxies in the universe are exceptionally rich and dense, far exceeding the large clusters near the Milky Way, such as the Virgo and Coma Clusters. Most of these giant clusters are very distant. Cl 2244–02, in the constellation of Aquarius, is one of the gigantic swarms of galaxies. Like all the large clusters it includes thousands of galaxies, but in this case most surround a group of a dozen giant ellipticals, closely huddled together in a sphere that is only 300 000 light-years across. Such a tremendous concentration of galaxies is very rare in the universe. There is not a single spiral galaxy at the heart of Cl 2244–02; all the galaxies of this type have been dismantled, stripped of their gas, and probably swallowed up by the most massive of the ellipticals billions of years ago.

Cl 2244–02 has a redshift $z = 0.33$. We observe it as it was when the universe was 75% of its current age, which represents a time shift of around 3 billion years. It is the splendid gravitational arc that surrounds it that made Cl 2244–02 famous. The enormous mass of the cluster curves space around it, as predicted by general relativity. The cluster, which is approximately spherical, behaves like an optical lens, concentrating and bringing to a focus the image of a distant background galaxy. The arc in Cl 2244–02 enables us to visualize the gravitational lens formed by the cluster. In fact, some years ago astronomers, using the theory of relativity's equations, developed a full theory of gravitational optics, so that subsequently, using measurements of the radius of curvature of gravitational arcs, they are able to calculate precisely the mass contained in the cluster within the so-called 'Einstein radius'. This method of 'weighing' clusters is absolute, because, by definition, no dark object, such as a cloud of gas, a cloud of dust, white dwarf, brown dwarfs, or black hole is able to escape. All the material, even if invisible, is involved in creating the gravitational lens. The apparent radius of the arc in Cl 2244–02 corresponds, for $z = 0.33$, and for an arbitrarily chosen cosmological model, to 150 000 light-years. For the region of the cluster corresponding to a sphere with this radius, the gravitational-optics equations give a formidable mass, amounting to 20 000 billion solar masses. This value, which is impressive in itself, reveals the exceptional density of Cl 2244–02. But this is not the main point: it is only when we compare the luminosity of the cluster with its mass that the

■ THE CLUSTER OF GALAXIES CL 2244–02, PHOTOGRAPHED BY THE HUBBLE SPACE TELESCOPE. THE CORE OF CL 2244–02 CONSISTS OF A DOZEN GIANT ELLIPTICAL GALAXIES, THE RESULT OF REPEATED GALACTIC COLLISIONS: NOT A SINGLE SPIRAL IS SEEN IN THE CLUSTER. THE SPECTACULAR GRAVITATIONAL ARC AROUND PART OF THE CLUSTER ENABLES US TO CALCULATE THE MASS CONTAINED INSIDE THE CIRCLE OF WHICH IT FORMS PART. THIS AMOUNTS TO SOME 20 000 BILLION SOLAR MASSES.

true significance of the latter becomes apparent. In fact, the total luminosity of the galaxies contained within the Einstein radius for the cluster Cl 2244–02 is around 200 billion solar masses. So we have 20 000 billion masses on one hand, and 200 billion on the other – the different methods of measuring the mass of the cluster disagree by a factor of 100. Cl 2244–02 is not the only cosmic curiosity of this sort: the same disagreement between measurements occurs with all clusters of galaxies.

This enigma, known as the problem of the missing mass, is not new. It has remained unanswered for some sixty years. It was first described in 1933 by the Swiss astronomer Fritz Zwicky, when he discovered strange behaviour in the Coma Cluster. The thousands of galaxies in this beautiful swarm showed abnormally high relative velocities. By comparing these velocities with the total mass of the cluster, estimated from the brightness of its individual galaxies, Zwicky discovered that the Coma Cluster should have long since dispersed. Its gravitational field appeared to be far too weak to hold onto its galaxies, which were moving at such high velocities. At that time, only one explanation appeared to be plausible to astronomers: the Coma Cluster was much more massive than its luminosity alone suggested. It is this invisible mass that must simultaneously cause the high velocities and also account for the cluster's coherence. The measure of the problem of the missing mass soon became evident. In calculating the rotational velocity of stars in the disk of spiral galaxies – such as the Milky Way – astronomers discovered that their behaviour was just as surprising as that of galaxies in clusters. Stars at large distances from the galactic centre were all revolving at a constant velocity – or even accelerating. In theory, far from the centre of mass, marked by

the galactic nucleus, these stars should slow down, as the planets in the Solar System do with increasing distance from the Sun. Once again, there was just one possible explanation: galaxies must be surrounded by a vast halo of invisible matter. In the case of the Milky Way, which is a typical large spiral galaxy, the ratio between the visible mass – condensed into stars and both bright and dark nebulae – and the hidden mass is about 1:10.

THE INTRIGUING COMPOSITION OF DARK MATTER

At the beginning of the 1980s, the consensus of opinion among astrophysicists was both simple and damning: they had no idea of the composition of most of the mass of the universe, amounting to 90% on a galactic scale and to 99% on the scale of clusters of galaxies. Yet they had done everything possible to detect this invisible matter. Even if they could not understand what it was, at least astronomers could state what this matter was not. Stars, even the extremely faint ones that had so far escaped detection, could not account for the missing mass. The cumulative light from hundreds of billions of them, even if they were very faint, would have been detected by our telescopes. The less conspicuous interstellar gas would also have been detected, either with telescopes when it is heated and re-emits light absorbed from stars, or with radiotelescopes, when, too cold to emit light, it still emits radio waves. The situation seemed to improve during the 1980s, with the launch of satellites that carried telescopes sensitive to X-rays, and which led to a spectacular discovery. Clusters of galaxies were found to be shrouded in vast haloes of gas, consisting of extremely tenuous ionized hydrogen, absolutely undetectable with conventional instrumentation. The origin of this X-ray-emitting gas has still not been definitely established. Part of it may simply come from the hydrogen created by the Big Bang, which has become concentrated in the gravitational wells that the clusters represent. Another portion of this gas probably comes from the galaxies themselves. In the richest clusters, such as Cl 2244–02 or the Coma Cluster, spiral galaxies, 10–30% of whose mass normally occurs as hydrogen in the form of nebulae, have disappeared. Researchers believe that the powerful gravitational waves created by interactions or collisions between galaxies have blown off the hydrogen into intergalactic space. It is thus this gas, visible only with X-ray telescopes, that the Einstein, ASCA and Rosat satellites detected. Surrounding the clusters, and extending out to millions of light-years, this gas, despite the fact that its pressure is almost zero – it contains just a few atoms per cubic metre and so amounts to an almost perfect vacuum – adds considerably to the mass of the clusters, because of the vast volume that it occupies. In the case of the Coma Cluster, whose visible mass amounts to 10 000 billion solar masses, the mass of this invisible hydrogen would exceed 50 000 billion solar masses!

However, the spectacular discovery of X-ray emitting gas has not resolved the enigma of the missing mass. In fact, the total mass of the Coma Cluster, estimated from the orbits of its galaxies, is close to one million billion solar masses. Since 1937, then, the question posed by Zwicky has remained unanswered. What does the hidden mass consist of? Astronomers continue to search for it, using more and more specialized methods. In the 1990s, new imaging technologies allowed observation of the innermost regions of the haloes of certain galaxies, which have not been recorded on the relatively insensitive photographic emulsions used previously. Very pale and diffuse, and of a reddish colour, these regions probably consist of old red dwarf stars, which had previously escaped detection. This new component of the galactic halo might amount to several tens of

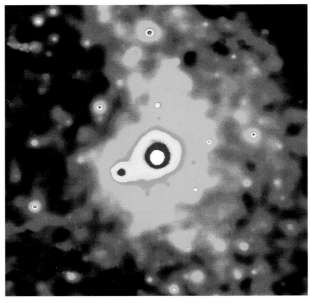

■ THIS IS AN IMAGE OF THE VIRGO CLUSTER, OBTAINED IN X-RAYS, WHICH MAY BE COMPARED WITH THE PHOTOGRAPH ON PAGE 100. AT THIS WAVELENGTH, THE GALAXIES IN THE CLUSTER ARE INVISIBLE, AND ONLY THE EXTRAORDINARILY TENUOUS GAS THAT PERMEATES THE CLUSTER IS REVEALED.

■ LIKE THE IMAGE OF THE VIRGO CLUSTER, THIS ONE OF THE COMA CLUSTER WAS MADE BY THE X-RAY TELESCOPE ON THE ROSAT SATELLITE. THE TOTAL MASS OF THE COMA CLUSTER IS AROUND 1 MILLION BILLION SOLAR MASSES: 20 TIMES AS MUCH AS ITS VISIBLE MASS.

billions of dwarf stars, most invisible, but would still represent just a few per cent of the overall mass of a galaxy like our own. This idea of invisible bodies has been actively explored for several years. Some theorists have indeed suggested that galactic haloes may be populated by millions of billions of tiny brown dwarfs, some of which may be no more massive than planets. Two methods have been adopted to find these extremely hypothetical objects. The first relies upon the relativistic effects that create gravitational lenses. If, in the course of its orbit around the galactic centre, one of these totally invisible, brown dwarf stars should chance to pass in front of a bright star, the former, acting as a gravitational lens, would briefly amplify the brightness of the latter, in accordance with the laws of gravitational optics.

By observing a large number of stars simultaneously, using telescopes fitted with extremely sensitive electronic cameras, brief stellar scintillations with very distinct characteristics should be detected frequently. Ambitious American and French experiments began in 1990 in Chile, Australia and France. Every night, small, wide-field telescopes monitor millions of stars not far from the Milky Way, in the Magellanic Clouds and in the Andromeda Galaxy. The very first results from this survey showed that tiny stars the size of planets probably do not exist in the universe. As for brown dwarfs, they also appear to be too rare in the halo to account for all the hidden mass. Theorists today believe that small, invisible stars increase the mass of the Galaxy by 10–20%, at most. Any hope of finding sufficient invisible stars in the galactic halo finally disappeared in 1995 and 1996 when the Hubble Space Telescope was also employed to search for the missing mass. Using that equipment, European and American astronomers carried out exceptionally accurate stellar surveys. The verdict from the teams led by Francesco Paresce and John Bahcall cannot be challenged: despite extremely deep 'probes' into the galactic halo, not even the Hubble Space Telescope has been able to find evidence for this hypothetical population of dwarf stars, even though, thanks to its extreme sensitivity, it was in principle capable of detecting them.

Over the years, all types of celestial objects have been invoked to try to account for this mysterious, invisible material. It now appears certain that neither dwarf stars – white or red – nor brown dwarfs, nor neutron stars, nor black holes can explain the missing mass. Such bodies would, in one way or another, leave tangible traces of their existence in

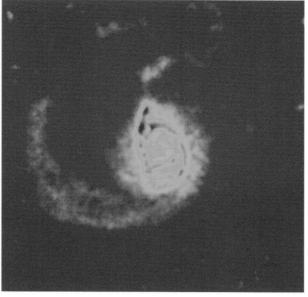

■ THIS SPECTACULAR RADIO IMAGE OF M 51 IN CANES VENATICI MAY BE COMPARED WITH THE ONE ON PAGE 107. AT A WAVELENGTH OF 21 CM, M 51 SHOWS A GIGANTIC EXTENSION TO ONE OF ITS SPIRAL ARMS, INVISIBLE ON CLASSICAL PHOTOGRAPHS.

the galactic halo and would, above all, have been detected by the equipment used to study gravitational lenses. There is, however, one last avenue of enquiry that astronomers can explore, and which has been suggested by Françoise Combes and Daniel Pfenniger. According to this Franco-Swiss team, the invisible material that everyone has been trying to find for so long may simply consist of an almost infinitely large number of tiny clouds of cold hydrogen. Each one of these globules, the size of the Solar System, would have a mass equivalent to that of Jupiter. At the extremely low temperature that prevails in intergalactic space – about –270 °C – this hydrogen is undetectable. This appealing hypothesis does, however, have one grave defect, because at present it is impossible to either confirm or refute it...

Failing to discover the nature of the hidden mass, researchers have turned to understanding how it is distributed in space. Although invisible, the dark matter makes itself felt through the relativistic effects that it creates. In the cluster Cl 2244–02, for example, it is responsible for producing the immense gravitational arc. Without the hidden 60–70% of its mass, the cluster would not cause sufficient curvature of space-time and would not show the resulting image. By carefully studying the most massive clusters, astronomers are now able to detect very subtle gravitational effects on the images of the background galaxies. Details of the gravitational arcs – their curvature and their orientation, for example – enable researchers to 'see' (rather

■ THE GALAXIES M 81 AND M 82 ARE BATHED IN A CLOUD OF HOT HYDROGEN, SEEN ONLY WITH A RADIOTELESCOPE. THE MISSING MASS MIGHT CONSIST OF TINY CLOUDS OF COLD HYDROGEN. NO INSTRUMENT EXISTS THAT WOULD BE CAPABLE OF DETECTING THEM.

as in a shadow play) the invisible mass that causes them. In the 1990s, astronomers were thus able to determine that the hidden mass envelops individual galaxies and, on a larger scale, also surrounds the central regions of clusters, just like the gas that emits X-rays.

If the enigma of the missing mass seems to motivate observers so strongly, it is because it has a far greater significance than just the desire to 'weigh' the galaxies. What is at stake is the most important question in cosmology.

OPEN AND CLOSED MODELS OF THE UNIVERSE

According to the Big-Bang theory – itself a result of general relativity – the universe is expanding from an infinitely dense and hot original state. General relativity is a theory of gravitation. It describes a curved space-time, whose geometry is controlled by the mass it contains. The study of the universe, for cosmologists, may be summarized as that of this expanding space-time. For the theoreticians, the receding galaxies are so many beacons that should, one day, allow them to determine the speed of the universe's expansion and its age, through the Hubble Constant. Yet according to the theory of relativity, not just the expansion of the universe, but also its geometry and, finally, its future, are entirely dependent on its mass. More precisely, the evolution of the universe is determined by its mean density – its mass per unit volume. This is the real significance of the search for the missing mass. Astronomers want to measure the density of the universe over a sufficiently large scale for it to be truly representative of its real density. The mean density, known as Ω, is of crucial importance in cosmology, in that models of the Big Bang differ radically depending on the numerical value assigned to it.

In the relativity equations, Ω has a value of 0, when the universe is reduced to a space-time that is in expansion, but which does not contain any mass. This model of the universe, reduced to its simplest form, is obviously not compatible with the real universe, even though the latter is, in fact, essentially empty. All models of the universe in

■ THE STATISTICAL STUDY OF THE DISTRIBUTION OF
GALAXIES LYING AT VERY GREAT DISTANCES – THROUGH
THE GEOMETRICAL DISTORTIONS CAUSED BY THEIR MASS
– ENABLES US TO EVALUATE THE DENSITY OF MATTER

which the density Ω lies between 0 and 1 are said to be 'open'. In these particular forms of the universe, which are described as being hyperbolic, space is infinite, the expansion is infinite, and the universe itself is eternal. Universes where the density is greater than 1 are described as being 'closed'. In these forms, the expansion of space-time eventually slows down, being constrained by the matter that it contains. After a certain time – several tens of billions of years in most models – the expansion ceases, and then the universe collapses! The galaxies approach closer and closer to one another, and instead of a Big Bang there is a Big Crunch, in which the whole history of billions upon billions of worlds abruptly plunges into oblivion.

In what sort of universe are we actually living? Space is, as we have seen, essentially empty; in round figures, intergalactic space contains, on average, one atom per cubic metre. This means that the universe has a density of the order of 10–30, one thousand billion, billion, billionth of that of water! In the cosmological equations, this figure – taking account of the missing mass, estimated from its gravitational effects – corresponds to a value of Ω that lies between 0.1 and 0.2, and implies that the universe has hyperbolic curvature, is infinite, and will expand for ever.

There remains one model that fascinates cosmologists, that where the value of Ω is exactly equal to 1. In such a universe, the mean density corresponds to a space-time that has infinite curvature, otherwise described as being 'flat', and agreeing precisely with the canons of geometrical perfection that Euclid defined twenty-three centuries ago. This Euclidean model of

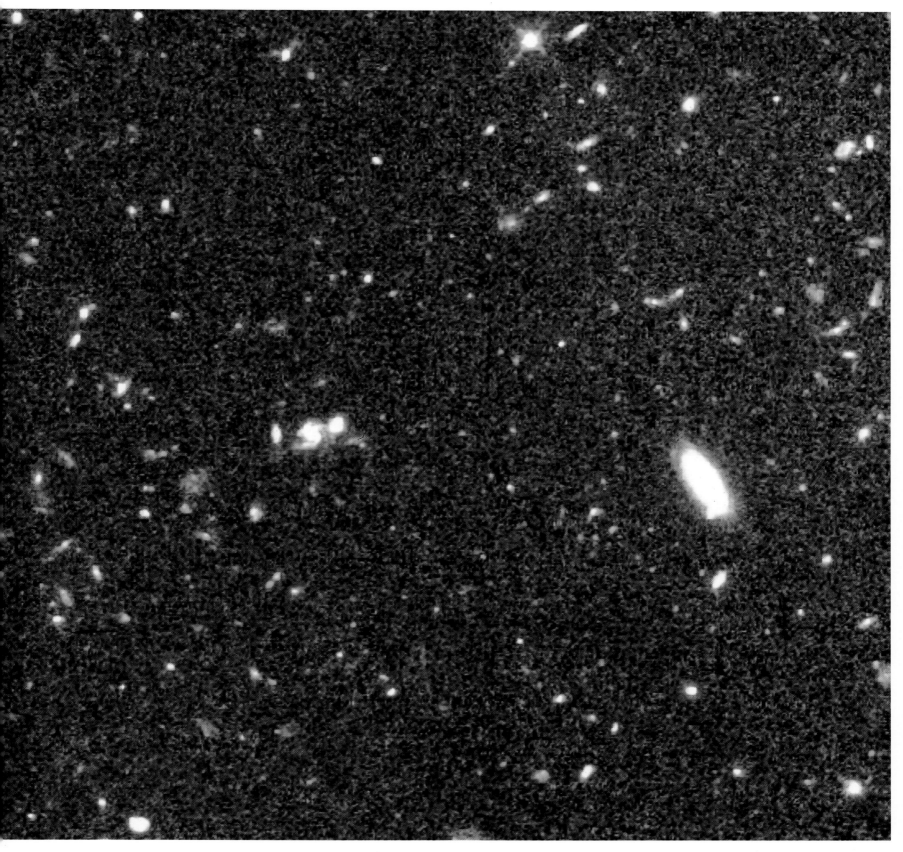

CONTAINED WITHIN A PARTICULAR VOLUME OF SPACE, AND THUS TO ESTIMATE THE CURVATURE OF THE UNIVERSE: THE IMPORTANT PARAMETER Ω.

the universe is probably the simplest and the most intuitive: space is infinite and is continually expanding. Put more precisely, in it the expansion is continuously slowing down and ceases after an infinite amount of time. For some fifteen years, one school of cosmologists, led by Alan Guth and Andrei Linde, has been studying the properties of this model of the universe, which is aesthetically extremely seductive. Using complex arguments in theoretical physics, they have been trying to convince the scientific community that it describes the real universe. But they have a difficult task. If the universe does obey this ideal geometry, they need to prove that it is ten times as dense as observations suggest, or else show, by a new interpretation of general relativity, that the classical theory of the Big Bang needs

to be radically altered. We shall return to this point in the last chapter.

This is where we are today. Whatever value of the density parameter is chosen by astronomers – $\Omega = 0.1$ or $\Omega = 1$ – and whatever geometry they adopt to represent the universe – a sort of hypersphere with three curved dimensions, or flat three-dimensional space – everyone is convinced that we live in an open, infinite, and eternal universe. ∎

The search for
the ultimate

■ SINCE THE BEGINNING OF THE DECADE, GIANT TELESCOPES ON THE
GROUND AND THE HUBBLE SPACE TELESCOPES IN ORBIT HAVE BEEN
PROBING THE UNIVERSE BACK INTO TIME TO AN EXTENT THAT
ASTRONOMERS NEVER DREAMED OF ATTAINING. THE HUBBLE DEEP FIELD
IS A TINY PATCH OF SKY IN URSA MAJOR, WHICH THE SPACE
TELESCOPES IMAGED OVER A PERIOD OF THIRTY HOURS. THIS
IMAGE SHOWS A SORT OF 'SECTION' OF SPACE-TIME, IN
WHICH DEEPER AND DEEPER STRATA OF THE UNIVERSE ARE
REVEALED.

■ A BLUISH STAR, LYING ABOUT ONE HUNDRED LIGHT-YEARS AWAY FROM THE EARTH, AND A DISTANT GIANT SPIRAL, WITH A REDSHIFT $z = 0.5$, CREATE AN EXTRAORDINARY COSMIC PERSPECTIVE. THE LOOK-BACK TIME TO THE GALAXY IS ACTUALLY 6 BILLION YEARS.

There are galaxies by the thousand, peppering a distant field of sky in the constellation of Ursa Major. A sky that is strangely devoid of stars. Just two of them, one bluish, the other orange, are visible in this dizzying cosmic landscape. By one of the paradoxes of astronomical imagery, the field is both minute – it corresponds to the tiny patch of sky visible through the eye of a needle held at arm's length – and also immense. It is, in fact, the deepest image of the universe that has ever been obtained. It is a fantastic, abstract cosmic landscape, consisting of four intertwined dimensions, three of space and one of time. This photograph will go down in the history of science as the first attempt by astronomers to include in a single image the entire history of the universe. To obtain this electronic image, known as the Hubble Deep Field, NASA monopolized the Space Telescope for more than a week, in December 1995. To be precise, the electronic camera's shutter remained open, trained on a spot in the constellation of Ursa Major, for 130 hours at a stretch. This region of the sky was carefully chosen by astronomers. The HST's line of sight was at right angles to the plane of the Milky Way, so that there were as few stars as possible. In fact, this area in Ursa Major is one of the most sparsely populated in the entire sky, which is an absolutely essential condition if one hopes to look far out into space and also back to the early stages in the history of the universe, without being inconvenienced by the Milky Way, nearby galaxies, and even slightly more distant clusters of galaxies. After five day's continuous exposure, the HST reached, for the first time, the incredible limiting magnitude of 30, which corresponds to the brightness of bodies that are ten billion times fainter than the faintest stars visible to the naked eye. The HST photographed more than 2000 galaxies in this field. That represents nearly 500 000 galaxies over an area of sky equivalent to that of the Full Moon. Finally, if the HST had photographed the whole sky under the same conditions – and without being affected by light from the stars – it would have recorded between 50 and 100 billion galaxies. This represents the limit for any current census of the number of galaxies in the universe. Because of the 130 hours of exposure for each field, however, it would have taken the HST some 370 000 years to carry out this comprehensive survey of the universe.

In relativistic cosmology, the Earth – like any other point in space – lies at the apparent centre of the universe. The point at

■ THE HUBBLE DEEP FIELD IS PROBABLY THE MOST EXTRAORDINARY RECORD EVER OBTAIN BY ASTRONOMERS. A FEW STARS — RECOGNIZABLE BY THEIR DIFFRACTION SPIKES — BELONGING TO OUR OWN GALAXY ARE SPARSELY SCATTERED ACROSS THIS FIELD IN URSA MAJOR, BUT MOST OF THE OTHER OBJECTS ARE EXTREMELY DISTANT GALAXIES. THE HUBBLE DEEP FIELD ENABLES US TO SEE MORE THAN 90% OF THE UNIVERSE'S HISTORY AT A SINGLE GLANCE.

■ THIS IS ONE OF THE MOST STRIKING IMAGES OF THE DISTANT UNIVERSE EVER OBTAINED. TAKEN WITH A TWELVE-HOUR EXPOSURE ON THE KECK II TELESCOPE, IT SHOWS THE RXJ 1716+67 CLUSTER OF GALAXIES, WITH A REDSHIFT $z = 0.83$. THOUSANDS OF GALAXIES ARE GATHERED IN THIS SCENE, WHICH IS 8 BILLION YEARS OLD.

which the universe originated, in the Big Bang, lies all around us, projected onto an inaccessible, fictitious sphere that lies at an infinite redshift. The 3-K background radiation, which was emitted 300 000 years after the Big Bang, is itself projected onto a sphere, lying at a redshift of $z = 1000$. In principle, then, a sufficiently powerful telescope could observe the entire history of the universe, from our own epoch back to practically the Big Bang, if it were aimed in some direction on the sky where there were successively greater redshifts, corresponding to epochs farther and farther in the past. Prior to the Hubble Deep Field, no astronomical instrument had reached as far out into the universe. There were many questions that bothered astronomers before this deep sounding was obtained: What does the distant universe look like? How far back in time can we go towards the Big-Bang singularity? Where, or more precisely, when were the very first galaxies formed? And finally: When did the first generation of stars appear?

These questions are not new. Throughout this century, every time a new giant telescope was commissioned, astronomers hoped to reveal the dawn of history, the first signs of the matter produced in the fiery heat of the Big Bang, followed by the development of large-scale structures in the universe. They hoped to unveil the formation of primordial galaxies, accompanied by the blaze of the very first generation of stars. Their ambition was that they would then be able to follow the evolution – here quiet, there violent – of these myriads of island universes, through billions of years, all the way from the Big Bang to the present day.

156

THE SEARCH FOR THE ORIGINAL GALAXIES

Cosmologists are all agreed that galaxies appear from infinitesimal fluctuations that occurred extremely early in the history of the universe. When the latter was nothing more than a searingly hot, expanding fluid, way beyond $z = 1000$, interactions between particles should have created gravitational instabilities, which started to grow. It was these instabilities, at different scales, that contained the seeds of galaxies, of clusters of galaxies, and of superclusters. With the expansion, the universe became less and less dense except in these local perturbations, which started to collapse upon themselves, creating vast, rotating, gaseous objects: embryo galaxies. Some specialists suggest that the first massive black holes formed at the centre of these protogalaxies, even before the stars, which would have appeared later. Others hold an opposing view, and believe that it was stars that brought an end to the universe's early 'dark ages'.

For a long time astronomers have, therefore, been hoping that they would be able to look back to embryo galaxies, which are fundamental pieces in the overall cosmological puzzle. Neither the famous telescope at Mount Palomar in the 1960s, nor the Keck Telescopes that were commissioned in the 1980s, nor even the Hubble Space Telescope at the end of the millennium, have been able to get anywhere near this crucial period in history. In fact, cosmologists, whose hands are tied, find that there is a true no-man's-land in the universe, which is forbidden to astronomers, and which extends from the 3-K background radiation – which was emitted about 300 000 years after the Big Bang, and has a redshift of $z = 1000$ – and the most distant objects that are seen between $z = 3$ and $z = 5$. Between these two ages, which – depending on the cosmological model used – are separated by 500 million to 2 billion years, what happened?

Probing the universe's distant past became possible only with the arrival of giant telescopes and the electronic cameras with which they are fitted. Until the beginning of the 1980s, telescopes were hardly able to reach back more than 5 billion years, i.e. to about a redshift $z = 0.5$. But the universe looked the same at that epoch as it does today. The same spiral and elliptical galaxies clump together into clusters identical to those found nowadays. As far back as that, the universe seems to obey the perfect cosmological principle; it is immutable, and always looks the same at all scales of both space and time. If the universe really does evolve, we have to look at an even earlier stage in its history to prove it.

At the beginning of the 1990s, the Franco-Canadian team of the astronomers Olivier le Fèvre and David Crampton undertook a systematic study – a sort of anticipation of the Hubble Deep Field – of the characteristics of all the galaxies contained within a narrow cone, centred on the Earth, i.e. at $z = 0$, and which, because of the instrumental limitations of the 3.6-m Canada-France-Hawaii Telescope, could only reach as far as $z = 1$. That was travelling back through 65% of the universe's history, which, for the first time, enabled us to see with our own eyes the evolution of galaxies similar to the Milky Way over a period of time amounting to 10 billion years – i.e. twice the age of the Sun – if the age of the universe itself is 15 billion years. The work of the Franco-Canadian team, subsequently taken up by other groups, who arrived at the same result, gave us our first view of the universe in that distant past. The astronomers discovered that at $z = 1$, the sky was far different from its current appearance. Spiral galaxies were more numerous, more luminous, and more active, the site of extraordinary bursts of star formation. Their spiral arms were full of enormous nebulae, glittering with millions of young blue supergiants. In addition, collisions between galaxies were far more common at that epoch than they are today. This was, of course, firstly because the scale of the universe $(1 + z)$ was only half what it is today, and also because galaxies were more numerous at that time. Astronomers now believe that the giant galaxies found today – including our own Milky Way – have formed from smaller galaxies that have coalesced over the course of time.

These deep probes back into the history of the universe have other objectives than just investigating the very first generation of stars and recreating the evolution of galaxies. Observers have several analytical methods

■ THE GIANT GALAXY HDF 9 HAS A REDSHIFT = 0.93, WITH A LOOK-BACK TIME OF 55-60%, DEPENDING ON WHICH MODEL OF THE UNIVERSE IS CHOSEN. THE LIGHT REACHING US FROM THIS OBJECT TRAVELLED FOR ABOUT 10 BILLION YEARS BEFORE IT WAS CAPTURED BY THE CAMERA ON BOARD THE HUBBLE SPACE TELESCOPE.

that allow them, in principle, to measure the universe's rate of expansion, and thus its age, and also to determine its geometry. In other words, to measure the values of the Hubble Constant H_0, and of the density parameter, Ω. To do this, however, astronomers need to be able to cut the data from these probes, which start as simple electronic images of the sky, into successive layers in both space and time. Finding out which of the myriad galaxies detected in the Hubble Deep Field is at a greater or lesser distance in space-time is a real headache. For the brightest galaxies in the field – often, of course, the nearest and thus the least interesting – astronomers were able to obtain a spectrum, directly determining the value of the redshift z. Unfortunately, for the faintest galaxies, those whose brightness is below magnitude 25, spectral analysis is impossible: current instruments are not sufficiently sensitive. The situation is therefore extremely frustrating for cosmologists: they are unable to study the faintest objects – often the most distant and thus the most interesting – in the Hubble Deep Field. More than 1500 galaxies, of the 2000-odd recorded in the Hubble Deep Field, whose brightness lies between $m = 25$ and $m = 30$, still remain beyond their grasp. They can see them perfectly well, but they know absolutely nothing about them: their distance, intrinsic brightness, and evolutionary state. Cosmologists are convinced that there are, in the Hubble Deep Field, galaxies with redshifts of 5, 6, or even 7, but no one knows how to recognize them. These bodies are being seen at look-back times never before reached, around 95%.

In mid-1999, for the first time, astronomers did manage to take a step in opening up the 'dark ages', by discovering a galaxy in Ursa Major with $z = 6.68$. The redshift of this galaxy is so great that the ultraviolet light it emitted in the past, at a wavelength of 121 nanometres, is now detectable in the infrared, at 930 nanometres. We see this galaxy as it was when the universe was just 5% of its current age. In other words, if the latter is 15 billion years old, we are seeing this distant object 750 million years after the Big Bang. Astronomers have never previously delved so far back into the past. No one knows at present when routine spectral analysis will be possible for objects that are so faint. For this reason, NASA and ESA have begun a feasibility study for the next-generation, giant space telescope – between 10 and 100 times as powerful as the HST – which might be launched around 2007.

IMAGES OF THE UNIVERSE'S DISTANT PAST

Despite this, spectral measurements made of a few hundred galaxies in the Hubble Deep Field, obtained with the two

largest telescopes in the world, on Hawaii, and indirect methods, which are essentially statistical, have provided researchers with evidence in support of certain cosmological models. In particular, it now seems that the period of most intense activity in the whole of the universe's history occurred between $z = 1$ and $z = 3$, that is (in a model where the age of the universe is 15 billion years), between 10 and 13 billion years ago. The galaxies in the Hubble Deep Field at that epoch appeared brighter, more turbulent, and more active.

All these galaxies have exceptional intrinsic luminosities. Some are surprised in the midst of coalescing with a neighbour, while others, from their chaotic shape and bluish colour, show that they are the sites of abundant star formation. In any case, all of them bear witness to the fact that at $z = 2$ or $z = 3$, the universe was more chaotic than it is today, and that encounters between galaxies were more frequent. (It is true that at $z = 3$, the dimensions of the universe were one quarter of what they are at present.) Finally, it appears that the morphology of galaxies was very different from what we know today. In fact, at very great distances, the HST does not find the modern distribution of galaxy types, namely 60% spirals, 20% lenticulars, 15% ellipticals, and 5% irregular galaxies. The

distant universe – apart from spirals with abnormal forms – shows more than 40% irregular galaxies. This fact, which obviously violates the perfect cosmological principle, spectacularly puts models in which the universe evolves, like the Big-Bang theory, in confrontation with those that are static, and in which it does not evolve.

Nevertheless, the images that Hubble obtained from this distant past should be treated with caution. Relativistic effects might gravely interfere with the interpretation of data from the Hubble Deep Field, as with those of different deep soundings obtained with other large telescopes. For example, the images of galaxies beyond redshifts of $z = 2$ or $z = 3$ cannot be directly compared with those of modern galaxies. Because of the cosmological spectral shift caused by the expansion of the universe, the image of the galaxy HDF 151, at $z = 3.18$, for example, although photographed by the HST in the visible region of the electromagnetic spectrum, actually shows the galaxy as it was radiating at the time, in the ultraviolet. In this spectral region, however, astronomers have a poor knowledge of the sky, because they are unable to carry out observations from the ground. To overcome this defective vision, one of the Space Shuttle Endeavour's missions in 1995 was devoted to ultraviolet photography of nearby galaxies, with the aim of being able to compare their images with those that the HST

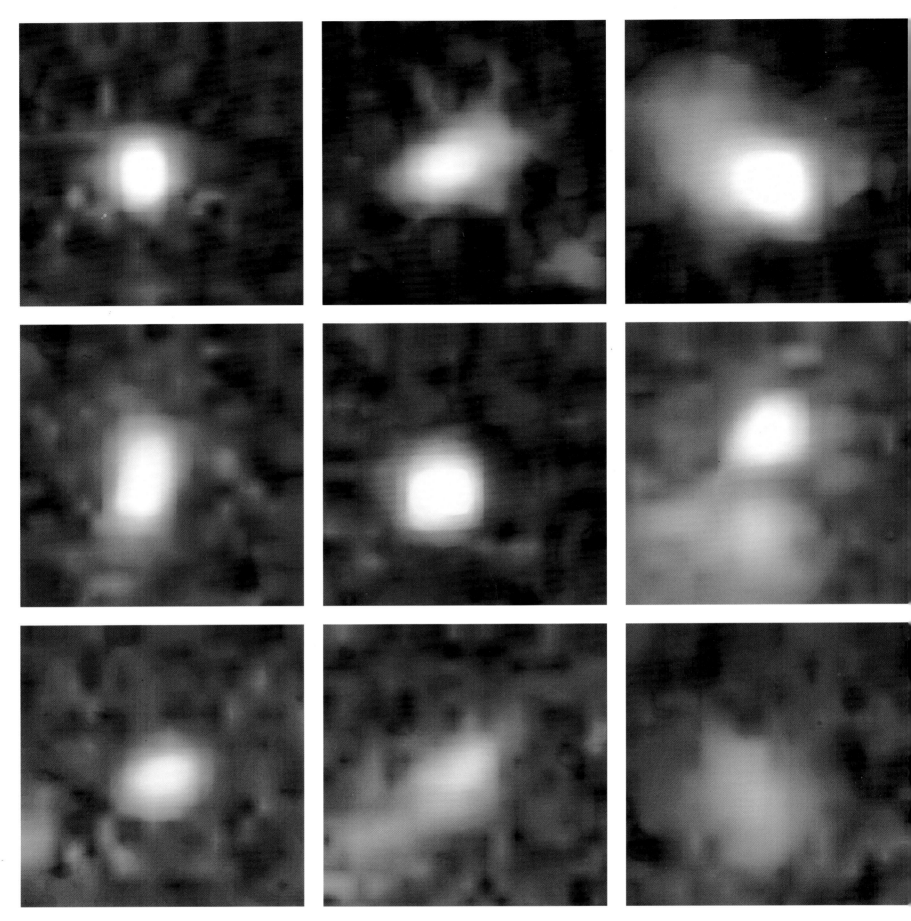

photographed at the farthest confines of the universe. These images are different from traditional optical images, because they are dominated by the youngest and hottest stars.

The distant universe revealed by the Hubble Deep Field does not just show violent and impressive galactic outbursts. Myriads of tiny, bluish compact galaxies, measuring no more than 2000 light-years in diameter, are also visible. These are the objects that astronomers suspect are the building blocks of modern-day giant galaxies, like the Milky Way. The latter, like all giant spirals, has arisen over the course of time, growing through incessant galactic mergers.

Quasars: primordial galaxies

■ These 18 objects were discovered in the constellation of Hercules by the Hubble Space telescope. All lie at z = 2.39 (which represents a look-back time of 80%), and are congregated into

The most distant objects in the universe have not, however, been discovered in the tiny field in Ursa Major, but more or less everywhere in the sky. These are the quasars. These objects, which appear like stars, are the extraordinarily brilliant nuclei of galaxies. They emit a violent flood of energy – quasars may shine as brightly as 100 to 1000 galaxies – probably because of the existence of a giant black hole deep inside them, which is continuously engulfing the surrounding interstellar gas. Heating of the gas as it falls into the black hole at a velocity that

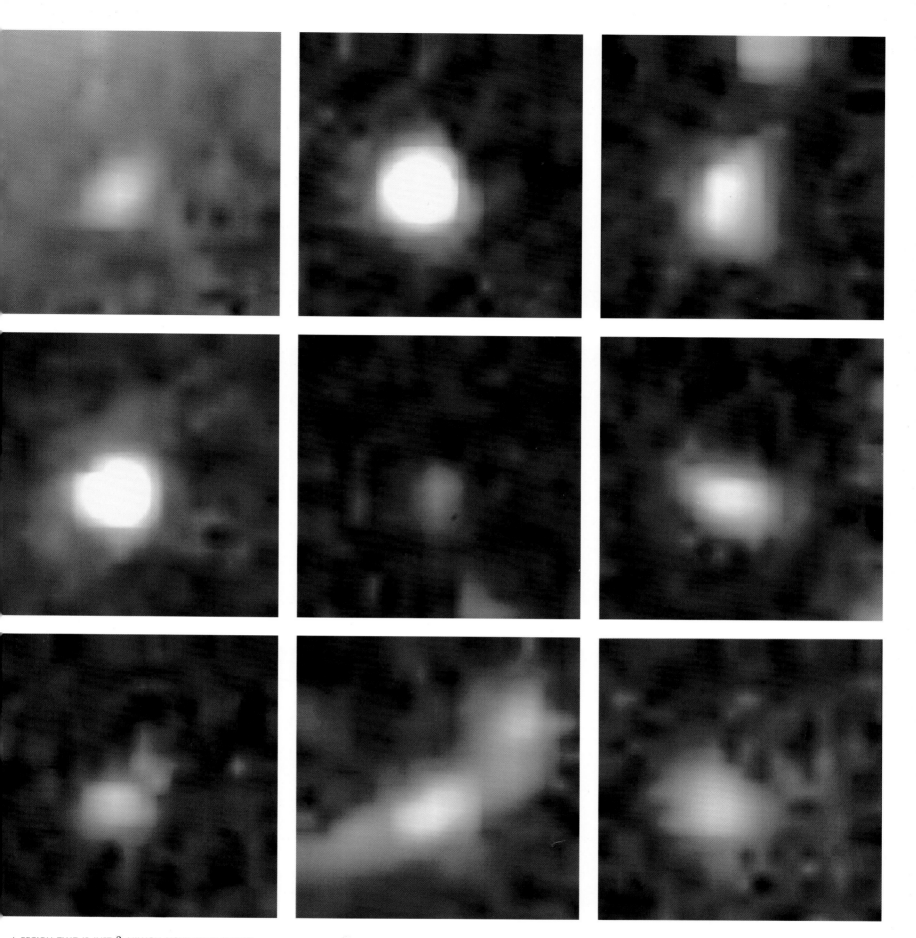

is close to the speed of light provides more energy than the nuclear processes occur-ring within stars. All giant galaxies may have passed through a quasar stage during the course of their evolution. At the centre of the Milky Way, for example, there is probably a black hole of more than 1 million solar masses, but this has largely swept the surrounding space clear of gas, and appears to be completely inactive at present. The giant galaxy M 87, on the other hand, has an intensely active nucleus, from which a gigantic jet of plasma is being ejected. There is probably a black hole of some 3 billion solar masses at the centre of the galaxy.

Far too faint to be classed as a quasar, this active nucleus does, nevertheless, represent quite closely what the quasar within a galaxy would be like – albeit some 100 times brighter. Since their discovery in 1963, quasars, because of their extraordinary brightness, serve as brilliant beacons, shining in the dark depths of space. For some time they have held all the distance records. PC1247+3406, discovered in the constellation of Canes Venatici in 1990, has a redshift of 4.9. It was not until 1999 that it was dethroned by a quasar with a redshift of 5.0. If the age of the universe is 15 billion years, these objects are being seen as they appeared more than 14 billion years ago.

The distribution of quasars has long intrigued astronomers.

Unlike 'normal' galaxies, they all lie at extremely great distances. The closest to us, the famous 3C 273, is at about 2 billion light-years ($z = 0.16$). This is an exception, because by far the majority of quasars lie much farther back in time, at redshifts with $z = 1$ to $z = 3$. Between that distant epoch and the present day, the number of quasars per normalized unit volume – i.e. taking the expansion of space into account – declines in a regular fashion. It was precisely over that range of redshifts that the activity of galaxies reached a violent peak, which leads us to think

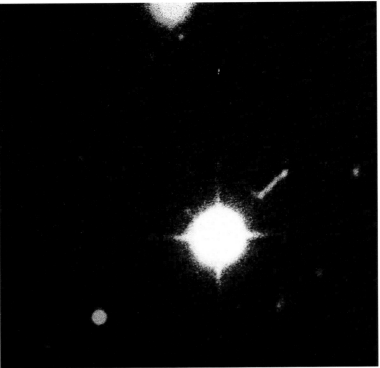

■ AT A DISTANCE OF JUST 2 BILLION LIGHT-YEARS, 3C 273 IS THE CLOSEST OF THE QUASARS. ITS EXTRAORDINARY LUMINOSITY IS GREATER THAN 100 GALAXIES LIKE OUR OWN. A JET OF PLASMA THAT IS ABOUT 100 000 LIGHT-YEARS IN LENGTH IS ESCAPING FROM 3C 273 AT SEVERAL THOUSAND KILOMETRES PER SECOND.

that quasars are born within the centres of galaxies when the latter either interact or fuse together. Currently, however, we only know 10 000 quasars in the whole universe, whereas there are tens of billions of galaxies. In the history of a giant galaxy, does the quasar stage pass extremely rapidly, and thus account for the extraordinary rarity of these objects?

How far back can we go in time, using the quasars as luminous beacons? Re-searchers soon discovered that at redshifts greater than $z = 3.0$, at which they are most numerous, the population of quasars de-clines the farther one goes back in time. Beyond $z = 4$, which represents a look-back time of about 90%, astronomers have found essentially none.

At the beginning of 1997, an international team led by the astronomer Peter Shaver published the surprising conclusion of their sys-tematic – and fruitless – search for distant quasars, using the radio telescope at Parkes in Australia, and the 3.6-m telescope at La Silla, in Chile. According to the researchers, the extremely powerful instrumentation of radio and optical telescopes

■ THE QUASAR PKS 2349 WAS DISCOVERED AT THE HEART OF A CLUSTER OF GALAXIES. IN THIS CASE, ASTRONOMERS SUSPECT THAT THE OBJECT IS THE RESULT OF A COLLISION BETWEEN TWO SPIRALS. PKS 2349 APPEARS HERE AS A BRILLIANT FOUR-POINTED STAR. THE POINTS ARE DIFFRACTION SPIKES CAUSED BY THE OPTICAL SYSTEM OF THE HUBBLE SPACE TELESCOPE.

that they were able to use should have easily enabled them to find bright quasars, among the thousands of objects whose redshifts they measured, even far in the past, at $z = 5, 6, 7$ or even more. According to Peter Shaver's team, if their instruments found nothing in the universe's distant past, it is simply because there was nothing to see. In other words, at that epoch, quasars did not yet exist. This discovery – or rather absence of discovery – confirms the Big-Bang theory, by offering new evidence of the evolution of the universe over the course of time. It also allows us, for the first time, to examine the epoch at which the very first giant black holes were formed, possibly at the heart of the first giant galaxies. This was somewhere between $z = 4$ and $z = 5$, which, if the universe is 15 billion years old, is 13.5 billion years ago, and an epoch strangely close to the Big Bang. Two or three decades ago, no astronomer would have imagined that anyone would ever observe such massive and extra-ordinarily luminous objects, fully formed, so close to 'time zero'. Yet all the current data seem to confirm that the universe really wa so surprisingly precocious. Quasars – which are too rare and too atypical to offer a reliable sample of the population of the distant universe – have been over-taken in recent years by normal galaxies, without giant active nuclei. Such bodies are currently being investigated in the Hubble Deep Field by teams of researchers who are trying to overcome the distance limit imposed by the sensitivity of spectrographs. These astronomers are using a new analytical method, which consists of recognizing

■ THIS EXCEPTIONAL PHOTOGRAPH TAKES US BACK TO THE UNIVERSE'S FAR DISTANT PAST. THE QUASAR Q000-263, IN THE CONSTELLATION OF SCULPTOR, ON THE LEFT, HAS A SPECTRAL SHIFT Z = 4.1, I.E. I,E, A LOOK-BACK TIME OF 90% ON THE RIGHT, IN THE FOREGROUND, IS A GALAXY AT Z = 3.3.

distant galaxies simply by observing their colour through specific filters. The Hubble Deep Field consists of four super-imposed images, obtained through ultraviolet, blue, red, and infrared filters, that have been combined to give the photographs that illustrate this chapter. Given that all galaxies have approximately the same characteristics as regards luminosity and colour, astronomers are searching for those that appear shifted towards the red. The statistical analysis of the Hubble Deep Field has thus enabled them to detect galaxies at redshifts of 5, 6 and 7. Unfortunately, without explicit spectral measure-ments, determination of the distances of these objects in space-time remains hypothetical.

Nevertheless, cosmologists are beginning to wonder about the physical processes that would have allowed the formation of galaxies from the homo-genous plasma of the Big Bang in just 750 million years (for a model in which the age of the universe is 15 billion years), or in an even shorter period (for models in which the universe is even younger). This situation is felt to be rather uncomfortable by some theoreticians, who also note that there is a serious inconsistency between the age of the universe as given by relativistic equations, on the one hand, and the age as measured from the observation of stars. In fact, for numerous cosmological models the age of the universe appears to be less than that of the oldest stars.

THE HUBBLE CONSTANT AND THE AGE OF THE UNIVERSE

This paradox concerning the age of the universe, which for certain researchers is the most serious challenge that has ever arisen to the tenets of the relativistic view of the universe, may perhaps be explained by the radically different methods that are used to date the universe. These are geometrical on the part of the cosmologists, and physical and chemical for astronomers who are trying to calculate how stars evolve.

In the simplest cos-mological models, where the density parameter (Ω) is 0.1, the age of the universe – also known as the Hubble Time – is defined as the reciprocal of the Hubble Constant, H_0. When $\Omega = 1$, on the other hand, the Hubble Time amounts to $2/3(1/H_0)$. All the models initially depend on the value adopted for the Hubble Constant. Measuring this to determine the age of the universe is one of the oldest projects in cosmology. But this measurement of H_0 is very tricky: it is a question of determining extremely accurately the distance of galaxies that are sufficiently far away, and then of measuring their apparent recession velocities. But the distances of galaxies, which are estimated by comparing the assumed intrinsic luminosity of their stars with the measured apparent magnitudes, were not known, until very recently, to better than a factor of 2. The objects that serve as 'standard candles' in galaxies are Cepheid variable stars (whose characteristics are well-known), supernovae, and finally, globular clusters. The maximum absolute magnitudes of the last two types of object are always approximately the same. In recent years, the Hubble Space Telescope has enabled astronomers to measure directly the magnitude of individual stars in the Virgo Cluster, even though they are some 50 million light-years

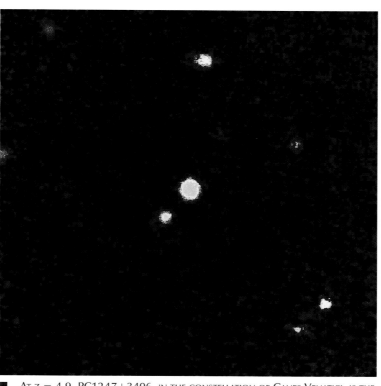

■ AT Z = 4.9, PC1247+3406, IN THE CONSTELLATION OF CANES VENATICI, IS THE SECOND MOST DISTANT QUASAR CURRENTLY KNOWN. PHOTOGRAPHED BY THE 10-M KECK TELESCOPE ON HAWAII, THIS BRILLIANT GALACTIC NUCLEUS APPEARS AS IT WAS WHEN THE UNIVERSE HAD LESS THAN 10% OF ITS CURRECNT AGE.

away. In addition, at the beginning of 1997, the same instrument managed to photograph globular clusters in the Coma Cluster, at a distance of more than 300 million light-years. At the beginning of 1997, using this set of measurements, astronomers estimated that the value of the Hubble Constant was about 75 km/s/Mpc. Put more simply, this means that two points in space, 3.26 million light-years apart, are separating at 75 km/s. Galaxies in the Virgo Cluster are thus receding at 1200 km/s, and those in the Coma Cluster at more than 7000 km/s. This is the rate at which the universe is expanding. It is as yet impossible to measure the value of the Hubble Constant in the past – although astronomers are already thinking about it! – but cosmologists still have to make allowance in their equations for a deceleration parameter, Q_0, which is gravity's gradual response to the essentially instantaneous release of energy in the Big Bang, and which tends to slow down the expansion over the course of time.

The age of the universe corresponding to a Hubble Constant (H_0) of 75 lies – according to the various cosmological models, i.e. depending on the values adopted for the deceleration parameter, Q_0, and the curvature parameter, Ω – between 9 and 13 billion years. Yet until recently, astrophysicists set the age of the oldest stars between 12 and 18 billion years, much to the cosmologists' disgust. Models of stellar nucleosynthesis are among the most accurate in the field of physics, so astronomers are reluctant to 'constrain' them too much, to lower the age of the stars. Until the spring of 1997, researchers did not know how to reconcile the discrepancies between these two branches of astrophysics – which, if not completely incompatible, overlapped only at the extremes of their ranges – and wondered if relativistic cosmology was not in the throes of a major crisis. Other astronomers, however, noted that although different, the two values for the age were extraordinarily close, and despite being para-doxical, confirmed the Big-Bang theory. There still remained the question of explaining the difference and, above all, of reconciling a young model universe with astronomical observations.

THE HIPPARCOS REVOLUTION

In May 1997, publication

of the results of the Hipparcos satellite by the European Space Agency may have enabled the problem of the two different ages of the universe to be resolved. During its mission, which lasted from 1989 to 1993, Hipparcos measured the distance and the true magnitude of around one million stars. Astronomers were able, for the first time, to calibrate accurately the distance of the Cepheids, the variable stars that serve as distance standards for measuring the Hubble Constant. To their great surprise, the European astronomers found that previously the distances of Cepheids had been systematically under-estimated. This result is of fundamental importance, because it means that, like the distances of the Cepheids, those of galaxies should be increased. As a corollary of this new calibration of the Cepheids, the Hubble Constant should be decreased, so that its value probably lies between 60 and 70 km/s/Mpc. The universe, after all, might well be 13 to 15 billion years old, and compatible with the age of the oldest stars. Especially as the Hipparcos measurements appear to indicate that until now, astronomers have also over-estimated the age of the oldest stars! In fact, by measuring the precise distances of the most ancient stars in the Galaxy, Hipparcos showed that they had been systematically under-estimated. That comes down to saying that these old stars have higher intrinsic magnitudes than astronomers thought. Hence their energy loss is greater than expected and, inevitably, their lives are shorter. The age of the oldest stars in the Galaxy is therefore around 11–13 billion years, a value that is finally in agreement with the age of the universe as estimated from the value of the Hubble Constant.

The deep probes undertaken in recent years, although offering a new vision of our past, have also brought their own crop of new problems. Even before the observations by the HST, the team led by Olivier Le Fèvre and David Crampton at the observatory of Hawaii noted a very disturbing fact: the data showed, without a shadow of doubt, that elliptical galaxies at a redshift $z = 1$ resemble those in the universe today. Yet elliptical galaxies are thought to be the oldest structures in the universe: devoid of gas, they contain just old stars, mostly red dwarfs and red giants. Yet at $z = 1$, astronomers are, according to the standard

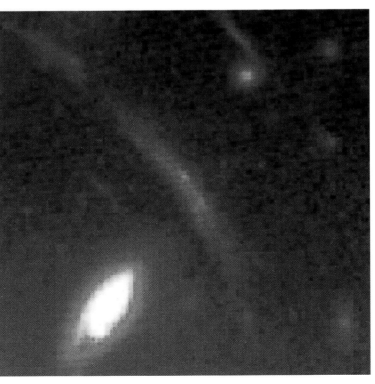

■ ASTRONOMERS HOPE TO DISCOVER THE MOST DISTANT OBJECTS IN THE UNIVERSE BY PROFITING FROM THE MAGNIFYING EFFECT CAUSED BY THE GRAVITATIONAL LENSES FORMED BY CLUSTERS OF GALAXIES. THIS GALAXY, AT $z = 2.5$, WAS DISCOVERED LYING BEHIND ABELL 2218 (*SEE* PAGES 140-141).

■ 4C41.17, SEEN HERE IN THE INFRARED WITH THE 10-M KECK TELESCOPE, IS ONE OF THE MOST DISTANT GALAXIES KNOWN. THE REDSHIFT OF THIS GALAXY AT THE ENDS OF THE UNIVERSE IS $z = 3.8$, WHICH AMOUNTS TO A LOOK-BACK TIME OF ABOUT 90%: 13.5 BILLION YEARS FOR A UNIVERSE THAT IS 15 BILLION YEARS OLD.

models of the Big Bang, looking 8–10 billion years into the past. How can we explain the existence of elliptical galaxies – themselves several billion years old – so early in the history of the universe? The discovery, in 1996, also at the observatory on Hawaii, of the galaxy 53W091 plunged astronomers into a sea of doubt. This galaxy in the constellation of Draco is currently the most distant elliptical known. At $z = 1.55$, the object has a look-back time of 70–75%. Yet spectral analysis of 53W091 shows that this galaxy largely consists of stars that are about 4 billion years old. This galaxy in Draco therefore puts cosmologists in a very awkward position, because finding a galaxy with such a redshift means that most of the models of the universe are null and void!

In a model where the age of the universe is 15 billion years, a redshift of $z = 1.55$ corresponds to a look-back time of nearly 11 billion years. To this we must add the galaxy's age of 4 billion years. Even taking the new data from Hipparcos into account, the total (15 billion years) takes us dangerously close to the Big Bang. When did this galaxy find the time to come into existence?

Observers and theore-ticians are currently exploring new avenues in an attempt to get out of this impasse. From the observational point of view, astronomers would greatly like to finally break through into the 'dark ages', which lie between $z = 5$ and $z = 1000$. This was the no-man's-land in which the universe's future destiny was determined, and which the

astronomers responsible for the Hubble Deep Field hoped to reach. But the HST's incredible sensitivity was not sufficient to cut through the mists of time. Statistical counts of galaxies in the Hubble Deep Field show that, even towards very low magnitudes – which means, in general, towards greater distances in space-time – the number of galaxies recorded does not decrease. This is proof, according to the specialists, that there is still something to see even farther out, and that the Hubble Space Telescope has not probed the full extent of the universe. Where are the first galaxies? When did the first generation of stars begin to shine? These are just two of the questions that will have to await the third millennium for answers.

In fact, numerical simulations show that if galaxies began to form less than 1 billion years after the Big Bang, they now lie at redshifts of between 10 and 15. Unfortunately, if they want to see such objects nowadays, astronomers need telescopes capable of observing at wavelengths between 4.4 and 4.6 μm –

that is, in the far infrared – because this is the region to which light from the galaxies' brightest stars has now been shifted. In addition, these galaxies in the farthest reaches of the universe cannot be any brighter than magnitude 30 or 31. At present no telescopes exist, either on the ground or in space, that are capable of detecting such faint infrared objects. That is, unless gravitational lenses are able to come to the astronomers' rescue once again, by allowing them to peer back over these vast distances of space-time. As we have seen, the richest clusters of galaxies form lenses, capable of amplifying the brightness and luminosity of galaxies in the background by 10 to 100 times. They behave just like natural telescopes. In 1996, a gravitational arc caused by the cluster Abell 2218, in Draco, was measured by the astronomer Jean-Paul Kneib and found to have a redshift of $z = 2.5$. This image of a distant galaxy has a look-back time of 75–80%, which

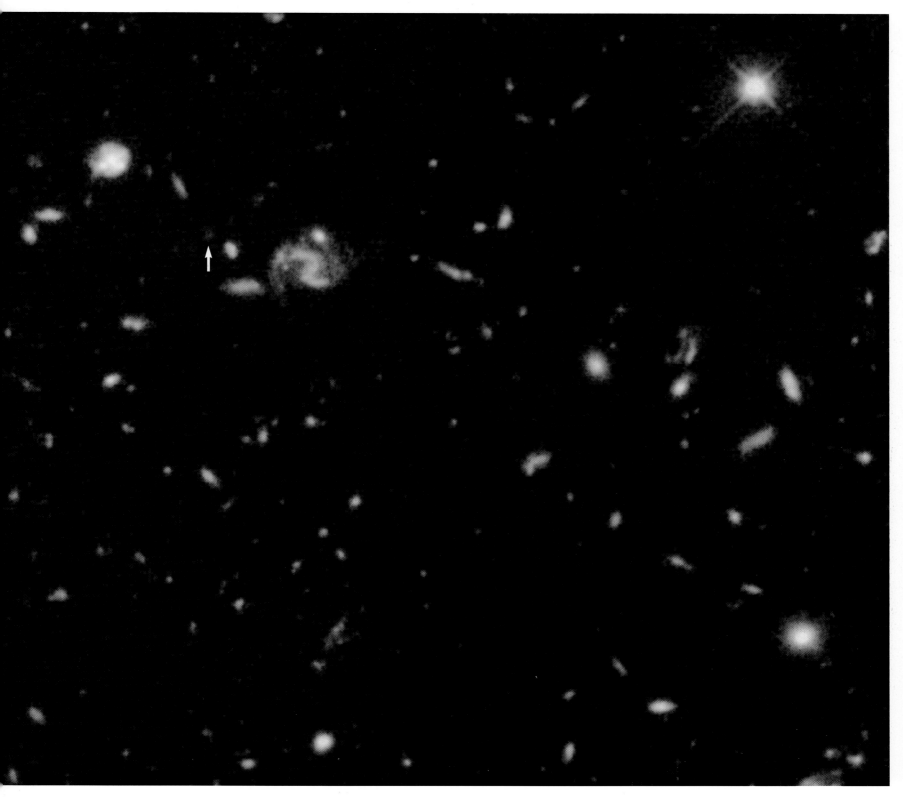

COLOURS OF GALAXIES, ENABLE RESEARCHERS TO ESTIMATE, IN AN INDIRECT MANNER, THE DISTANCES OF SUCH GALAXIES. THIS EXTREMELY RED GALAXY MAY, ACCORDING TO SOME ASTRONOMERS, BE THE MOST DISTANT OBJECT IN THE UNIVERSE, AT $z = 6$ OR 7.

means that, in a universe with an age of 15 billion years, we are seeing the galaxy as it was some 12 billion years ago. Despite its enormous distance, it is perfectly visible, and its spectral characteristics are easy to study. The cluster Abell 2218 has, in effect, increased the magnitude of the image of this distant object by about 3, i.e. it has amplified it about 15 times. In this same Draco cluster, and also in other giant clusters like Cl 1358+62, Abell 370, and Abell 2390, other, far more distant galaxies are waiting to be observed and measured by researchers. It is perhaps here, in these rich fields, which appear nearby because of the magnifying effects of the curvature of space, that astronomers will one day discover the very first galaxies. ■

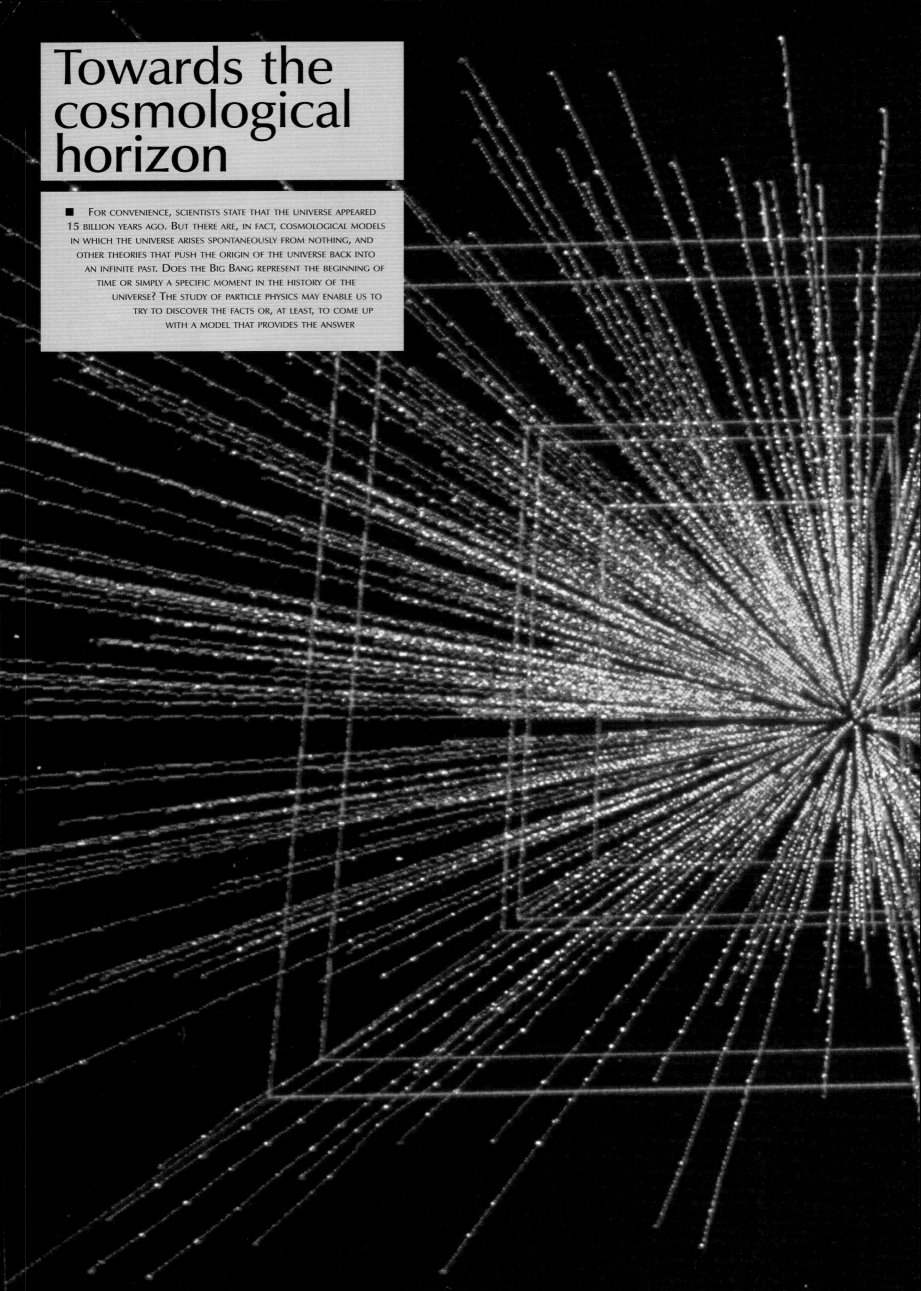

Towards the cosmological horizon

■ For convenience, scientists state that the universe appeared 15 billion years ago. But there are, in fact, cosmological models in which the universe arises spontaneously from nothing, and other theories that push the origin of the universe back into an infinite past. Does the Big Bang represent the beginning of time or simply a specific moment in the history of the universe? The study of particle physics may enable us to try to discover the facts or, at least, to come up with a model that provides the answer

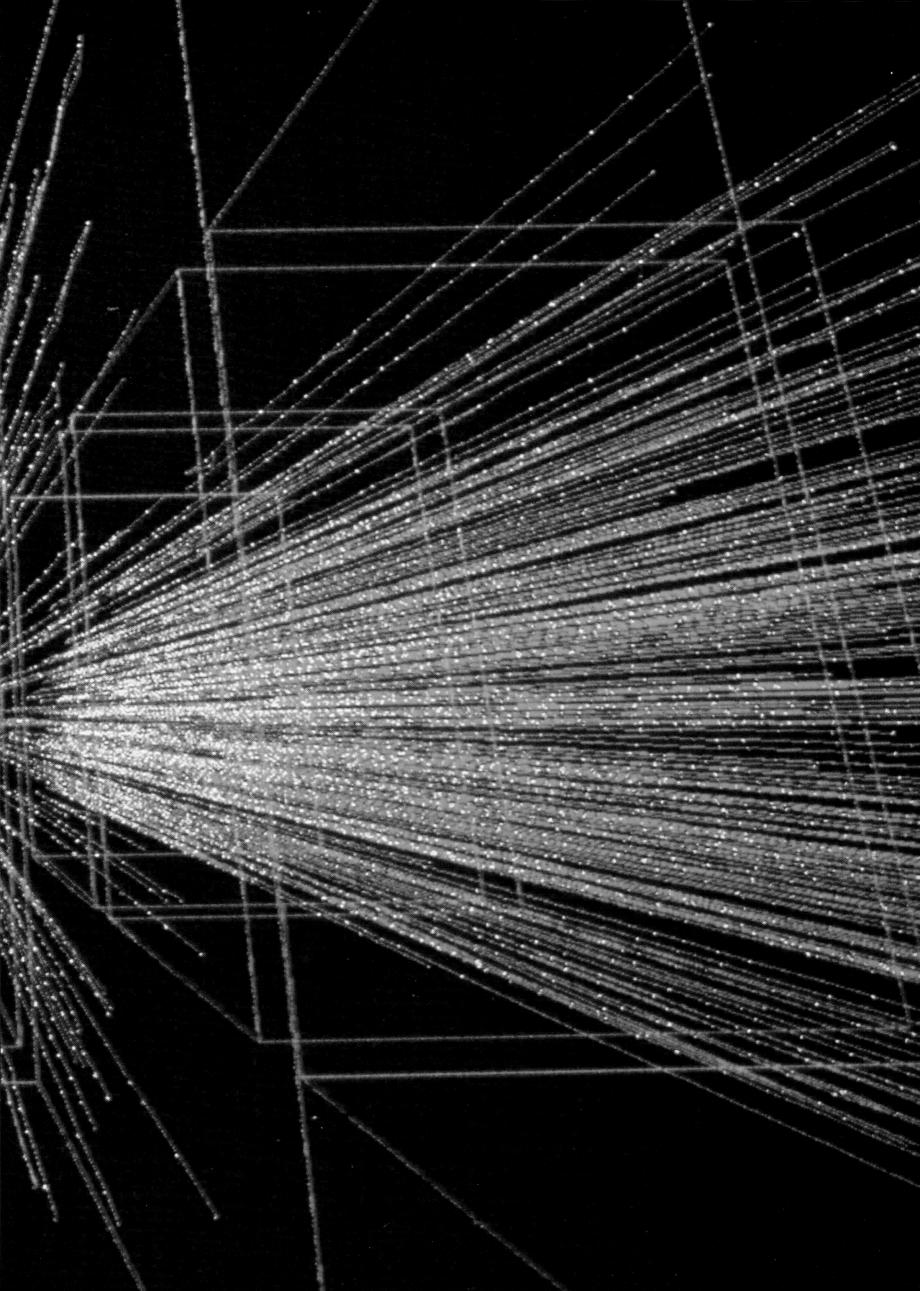

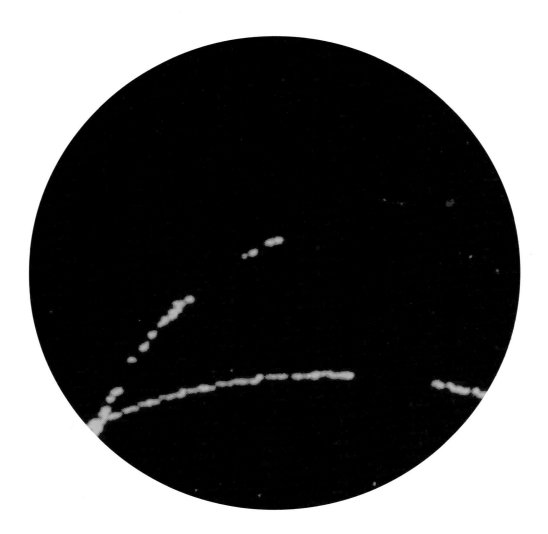

How should we visualize the Big Bang? Suddenly, out of some incomprehensible conditions, the universe appeared. Physicists try to imagine what happened 'before' time 10^{-43} s, which is our current limit of knowledge. But our notions of space and time probably have no significance for the universe at such a distant past.

It is the very first moment in time. The whole universe is contained within a single point, but a point with infinite dimensions... It is a sea of energy that has erupted from a dark, chaotic abyss. The Big Bang is a formidable challenge to human intelligence. This period of time escapes our understanding because it does not obey the principle of causality: it defies all attempt at description, whether mathematical or symbolic.

But what is the Big Bang – that explosive origin of the cosmos, that initial *Fiat lux!*, as it is often described in astronomical books? Does it really mark the beginning of history? Is it simply a single episode in the evolution of the universe, an instant in an infinite span of time? The stark truth is that no one knows. To comprehend what the Big Bang represents, it may perhaps be best to first understand what it is not. The Big Bang is not the explosion, in an infinite empty space, of a primordial atom in which the whole of the universe was concentrated. Above all, the Big Bang does not mark, or rather does not necessarily mark, the birth of the universe. Such simplistic, reductionist, and completely false representations have done a lot of damage to a theory that does

not deserve it. Nevertheless, the intuitive image, which is somewhat Newtonian, of a universe suddenly bursting into an empty, external, and pre-existent space-time still persists. In particular, it evokes the naive question: One day, will giant telescopes see the point on the sky where the Big Bang took place?

In the equations of space-time, the Big Bang marks the origin of space-time itself, and of all the energy that it contains. As a result, there is no single point: the Big Bang took place everywhere. If space is currently infinite, as certain cosmological models suggest, then it was the same at the time of the Big Bang. It was simply that at that epoch, any specific measurement of the universe, let us say 1 billion light-years, was confined to an infinitely tiny space. That goes against common sense, but is perfectly justified in the context of the equations of relativity. The dilation of space over the course of time; and the decrease in density and temperature: these are the only precise, and probably correct, images that cosmologists are able to give of the origin of the universe. By force of habit, for convenience, and because we do not know how else to express it, the Big Bang is always represented as

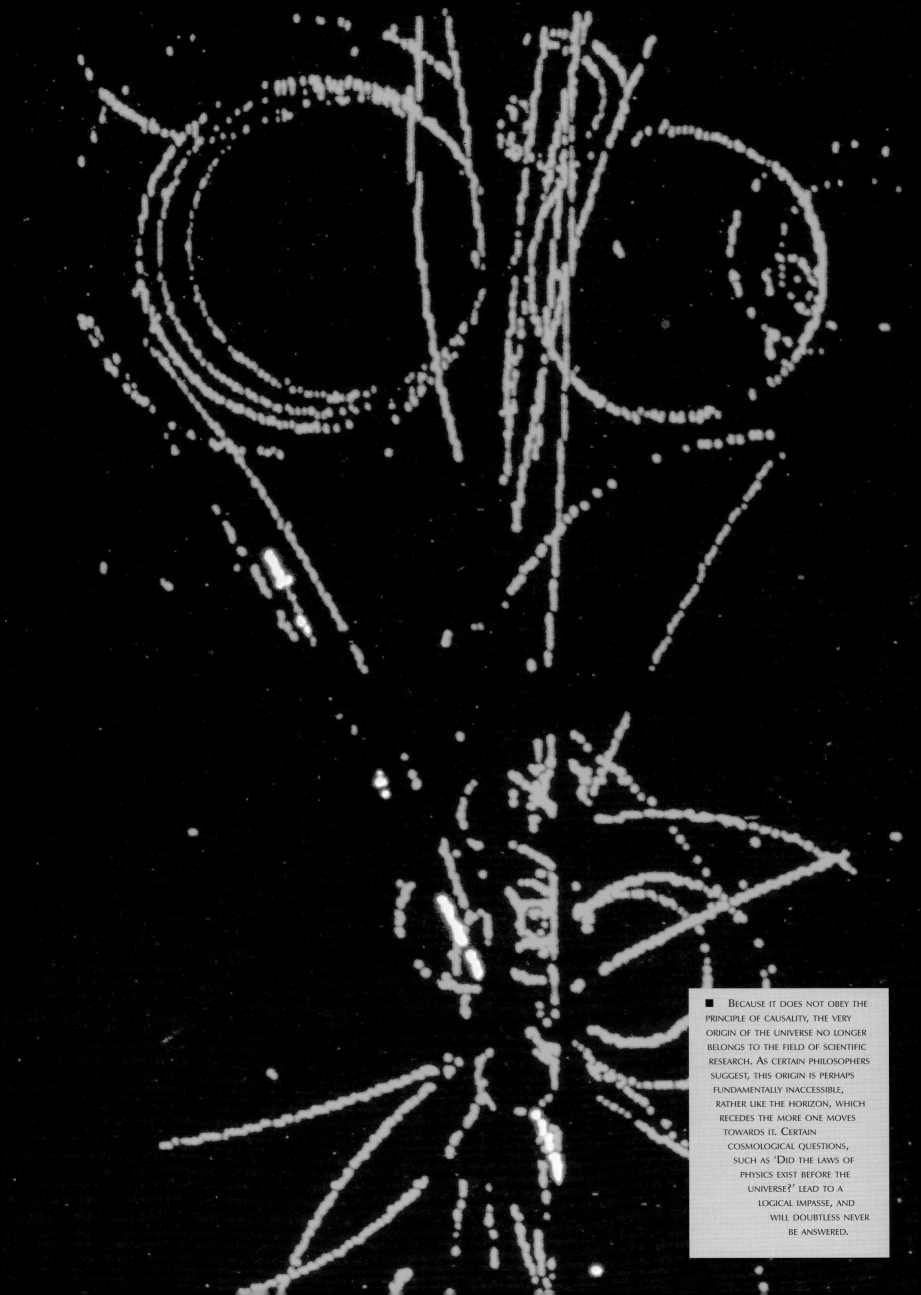

■ BECAUSE IT DOES NOT OBEY THE PRINCIPLE OF CAUSALITY, THE VERY ORIGIN OF THE UNIVERSE NO LONGER BELONGS TO THE FIELD OF SCIENTIFIC RESEARCH. AS CERTAIN PHILOSOPHERS SUGGEST, THIS ORIGIN IS PERHAPS FUNDAMENTALLY INACCESSIBLE, RATHER LIKE THE HORIZON, WHICH RECEDES THE MORE ONE MOVES TOWARDS IT. CERTAIN COSMOLOGICAL QUESTIONS, SUCH AS 'DID THE LAWS OF PHYSICS EXIST BEFORE THE UNIVERSE?' LEAD TO A LOGICAL IMPASSE, AND WILL DOUBTLESS NEVER BE ANSWERED.

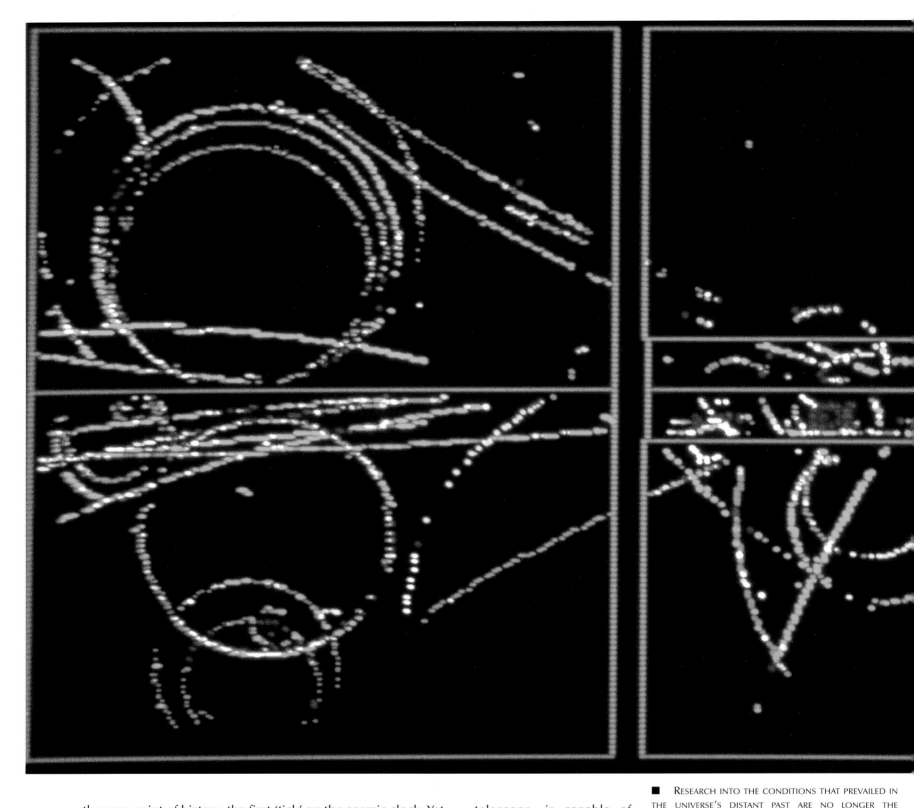

the zero point of history, the first 'tick' on the cosmic clock. Yet, as we shall see later, this scenario distorts a physical reality that is far more subtle but also more difficult to understand.

Cosmology aims to describe the universe as a whole, that is to say both its current structures as well as their evolution throughout the course of time. All modern astrophysics is sustained by the equations of relativity, which give numerical solutions to models of the universe, and endow them with a greater or lesser degree of probability. We may possibly be actually living in one of the theoretical models proposed by cosmologists, but possibly not.

The fundamental problem that the Big Bang poses for cosmology is that it actually lies outside the grasp of the theories. In principle, cosmology, sustained by observations obtained through the

telescope, is capable of moving back in time, following its own Ariadne's thread – the trace that light has left in space-time – to the very beginning of our past. When astronomers announce that they have observed a quasar at a redshift of $z = 4.9$, they evoke the distant past: they are looking back over 90% of the universe's past. At $z = 1000$, by observing the 3-K background radiation, they are able to detect the distant glow from the Big Bang, emitted just 300 000 years after the hypothetical time zero. It is actually the last echo of the Big Bang. Everything that happened before that period is inaccessible to our instruments, and will probably remain so for several decades. The equations of relativity, on the other hand, allow us to predict without great

■ RESEARCH INTO THE CONDITIONS THAT PREVAILED IN THE UNIVERSE'S DISTANT PAST ARE NO LONGER THE DOMAIN OF ASTRONOMY. THE UNIVERSE, IMMEDIATELY

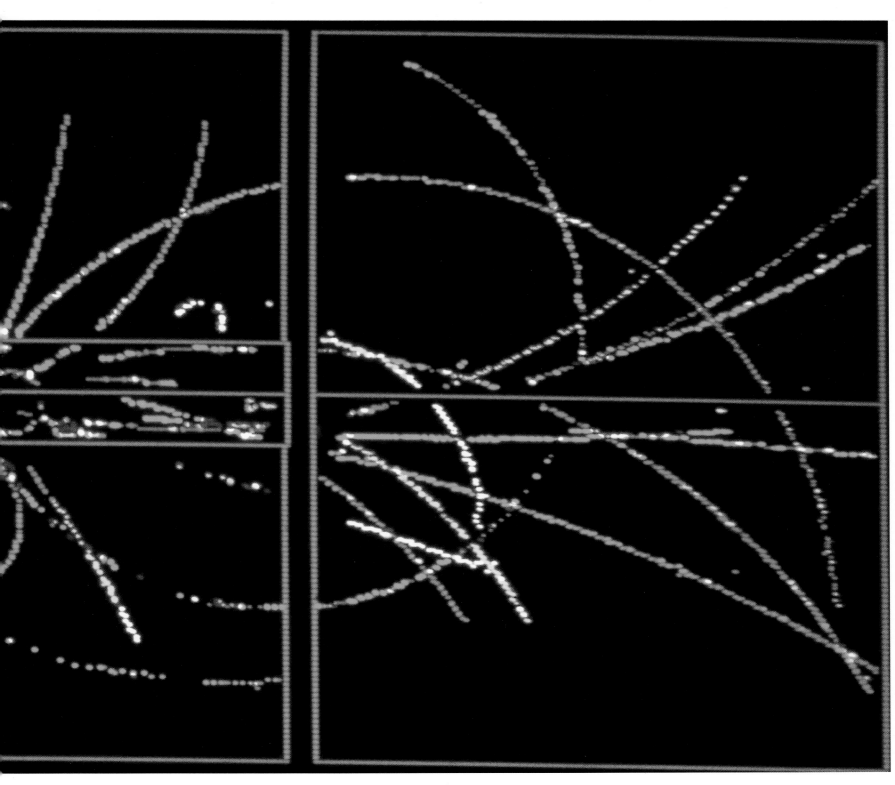

progressive increase in temperature and density as we move backwards towards the Big Bang have been calculated. But how far back in time can the physicists go? Some 300 000 years after the Big Bang, as we have seen, the temperature of the universal 'soup' was around 3000 K: that is the temperature of the surface of a star. Much earlier, about a quarter of an hour after the Big Bang, the temperature was several hundreds of millions of degrees. A temperature that is still common in the present-day universe, and that is attained in the cores of supergiant stars. But how can anyone imagine the whole universe as being like the searing plasma at the heart of Rigel or Deneb? Even earlier, around the first one thousandth of a second, the difficulty what happened 'before' that time. The

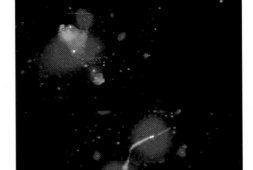

■ THE CLUSTER OF GALAXIES PKS 2104–25. THIS FALSE-COLOUR PICTURE WAS CREATED FROM A RADIO IMAGE, IN GREEN, AND PHOTOGRAPHS TAKEN IN VISIBLE LIGHT, IN BLUE AND RED.

temperature of the universe was close to 1000 billion degrees. A figure that may still be acceptable to physicists, but which cannot be compared with anything that exists in the universe today. Physicists are able to describe the extraordinarily dense and hot universe, back to some incredibly distant epoch in time. The behaviour of this plasma, and of the particles of which it was formed, may be predicted by their equations and even, in certain cases, reproduced in the laboratory, in particle accelerators. The universe, around the time of the Big Bang, is no longer studied with a telescope, but with the tools for dealing with the infinitely small.

By going back in time, physicists also discover a universe that becomes simpler and more unified. Today, four fundamental forces govern all physical phenomena. The strong nuclear force and the weak nuclear force govern the stability of atomic nuclei. These two forces have a limited range of action, essentially restricted to the dimensions of atoms. The electromagnetic force, by contrast, has an unlimited range, and is responsible for the behaviour of the electrons that encircle atomic nuclei, and for the reactions between atoms within molecules. Finally, gravity, which is very weak – the strong nuclear force is 1000 billion, billion, billion, billion times as strong – also has an infinite range, and governs the large-scale structure of the universe.

Physicists believe that these forces are derived from a single fundamental force that arose directly from the Big Bang, and which could account for all physical phenomena. The unification of the electromagnetic force and the weak nuclear force into the electroweak force occurs spontaneously in nature at a temperature of 10^{15} K. That is the temperature that the universe reached one thousand billionth of a second after the Big Bang. The prediction of the unification of these two forces by Steven Weinberg, Abdus Salam, and Sheldon Glashow, confirmed by the particle accelerator at CERN, earned them the Nobel Prize for Physics in 1979. In contrast, however, it seems that no present or future accelerator will be able to verify experimentally the unification of the electroweak force and the strong nuclear force into a single force, that has been called the electronuclear force, around 10^{28} K. In the history of the universe, the fusion of these three present-day forces occurred spontaneously before 10^{-35} s. There remains gravity, which physicists wisely assume could have been unified with the electronuclear force even earlier in the universe's history, during the era, known as the Planck era, which extended from the Big Bang until just 10^{-43} s.

This infinitesimally short instant is the final limit to time that is known, on the long road that has taken us back towards the Big Bang. At that epoch, 10 billion light-years of our present-day universe were contained within a volume just 10^{-33} m across, at a temperature of 10^{32} K, and at the apparently impossible density of 10^{94}. How should one imagine such a tiny volume of space? There is a greater difference in scale between an atomic nucleus and a size of 10^{-33} m than there is between the size of the present-day

universe and that same atomic nucleus! These values, although they seem mind-boggling or even downright impossible, are, in fact, accepted as such by the cosmological community, to whom they do still have some physical significance.

RELATIVITY AND QUANTUM MECHANICS

But before that? Physicists are currently unable to penetrate the Planck era, because their equations are not able to describe what took place within this 'Terra Incognita' in the primordial universe.

In fact, relativistic cosmology comes up against the very structure of relativity itself. The latter describes space-time as a sort of continuous fabric: in the relativistic universe, for example, a plane is as perfectly plane as it is in the imagination of geometers. In fact, when the cosmological equations trace the steam of time backwards towards the Big Bang – i.e. when they describe the increase in the density and temperature as a function of the decrease in the scale of space – they take Einstein's work to its idealised limit. At time zero, according to the equations, the universe had no dimensions, and its temperature and density were both infinite. The Big Bang therefore presents scientists with a mathematical singularity, a sort of scientific no-go area.

In reality, the theory of relativity is not describing the primordial conditions at this point, but is instead showing its own limitations. Physicists do not, in fact, accept the infinite values given by relativity's numerical solutions, and have discovered that at extremely small scales, Einstein's theory is no longer valid. Nowadays, physicists describe the realm of atoms by using another conceptual tool, the theory of quantum mechanics, which has a profoundly different mathematical structure from that of relativity. In the quantum world, the microcosmos, which it is quite impossible to comprehend on our own scale, is governed by Heisenberg's uncertainty principle, according to which sudden, unpredictable fluctuations appear spontaneously at extremely small scales, and which, in particular, prevent us from specifying the exact position of a particle in space. Unlike the continuity implicit in

■ THIS IS THE MOST DISTANT GALAXY CURRENTLY KNOWN IN THE UNIVERSE. DISCOVERED IN 1998 IN URSA MAJOR BY THE HST AND THE KECK II TELESCOPE, HDF4–473.0 HOLDS THE REDSHIFT RECORD WITH $z = 5.6$, AND TAKES US BACK THROUGH MORE THAN 90% OF THE HISTORY OF THE UNIVERSE.

relativity, the theory of quantum mechanics suggests that atomic structure is discontinuous. As against the precise location of particles in curved space-time, to which they are intimately linked, that relativity specifies, quantum theory gives a probable location – i.e. one that is intrinsically uncertain – for particles in a static and passive space-time that is, in certain respects, Newtonian. Although they perfectly complement one another when it comes to deciphering the two extremes of the present-day universe, these two theories run counter to one another and are mutually exclusive when it comes to describing the Planck era. Although relativity is no help when it comes to understanding the infinitely small, quantum mechanics does not take the spatial and temporal structure of the universe into account.

Following in the footsteps of Einstein, who, to the very end of his life searched in vain for an overall unification theory, cosmologists remain positive that they can reconcile these two profoundly different fields of physics. They are exploring new theories that will allow them to integrate the essential concepts of quantum mechanics and general relativity. Quantum gravity is just one of these new theoretical approaches. One of its strangest and most fascinating properties is that it postulates that space-time itself has a structure. If cosmologists have so much difficulty in understanding what happens during the Planck era, it is, according the quantum gravity theorists, because at this scale in the universe, relativity's continuous space-time breaks down, and like particles, is subject to erratic fluctuations in the quantum-mechanical regime. In some respects, seen from a distance and on our own scale, space-time appears to be perfectly smooth, calm – and curved – like the ocean seen from a plane, but, seen from very close by, on the quantum scale, it reveals that it is more like foam, blown about by fierce winds over rough seas.

One of the newest and most promising attempts to unify relativity and quantum mechanics is superstring theory, which proposes a revolutionary and somewhat disconcerting image of the infinitely small. In this theory, elementary particles, which physicists have, until now, visualized as infinitely small, dimensionless points, should be regarded as infinitely thin,

minute strings, with lengths that do not exceed 10^{-35} m. This theory is central to all current speculations, because its formal mathematics integrate relativity and quantum mechanics in a perfectly normal manner. It also allows the unification of the four fundamental forces and, finally, predicts the overall characteristics of elementary particles. In this strange world, particles are no more than different modes of vibration of these infinitely small strings. Superstring theory, whose mathematical beauty and internal consistency fascinate physicists, suffers from just one fault: it cannot be verified. The levels of energy that are required to test the existence of the unified field and the reality of strings are unattainable by particle accelerators, and thus render the theory beyond experimental proof.

Nevertheless, in the light of superstring theory and of other variations on the theme of quantum gravity, the unimaginable conditions of the Big Bang take on a new and extremely strange dimension. In the Planck era, both chronology and the principle of causality disappear in a chaotic, fluctuating space-time. It is no longer possible to time events, and the origin of the universe becomes inaccessible. Searching for the mythical time zero now has no more meaning for scientists than asking 'What came before the Big Bang?' In the theory of quantum gravity, space and time simply emerge from the Planck era, at around the time of 10^{-43}s. The cosmos emerged from 'somewhere else', where time did not pass, and time zero did not exist – which is another way of saying that the Big Bang never took place.

Although the Big Bang still seems to escape from scientific methodology – if only because with its unique status as an event without a cause, it violates the principle of causality – cosmologists have by no means finished wondering about the structure of the universe, which poses new problems as fast as there is any theoretical or observational progress. Apart from the enigma posed by the missing mass, there are two other mysteries that are just as fascinating, the cosmic horizon, and the curvature of space, both of which are tormenting the theoreticians. The latter have recently realised that the classic theory of the Big Bang

■ THE FAINT RED PATCH IN THE CENTRE OF THE PHOTOGRAPH IS 8C 1435+63, ONE OF THE MOST DISTANT GALAXIES KNOWN, WITH $z = 4.2$. THIS COMPOSITE IMAGE, TAKEN IN THE VISIBLE AND INFRARED REGIONS, WAS OBTAINED WITH THE 10-M KECK TELESCOPE AT THE OBSERVATORY ON MAUNA KEA.

is incapable of accounting for the appearance of the universe as we observe it today. According to astronomers, the sky appears identical in all directions. Everywhere we look, whether north, south, east or west, the same sort of clusters of galaxies are scattered across the sky, out to distances of several billion light-years. Similarly, the cosmological background radiation, emitted when the universe was just 300 000 years old, appears isotropic across the sky. It is this homogeneity that intrigues scientists. They point out that in the universe's very distant past, those regions from which we are receiving radiation today were never in close contact. Information, in the universe, propagates at the speed of light. An object indicates its existence by the radiation that it emits, which, like electromagnetic radiation, propagates at 300 000 km/s. At every moment, our vision of the world is thus limited by a horizon, which is expanding, ever since the Big Bang, at 300 000 km/s. So, one second after the Big Bang, the horizon for any individual point in the universe lay at a distance of 300 000 km. This cosmic horizon moves infinitely faster than the expansion of the universe. Which is another way of saying that with every second that passes new regions of the universe appear that were never connected in the past. Yet present-day cosmic structures show that there must have been an extraordinary homogeneity in the past, that cannot be expressed in equations. How could different regions of the universe, never in contact with one another, 'pass the word' to one another so that their characteristics were exactly the same at the instant that their cosmic horizons coincided, and they came in contact? For physicists, until some twenty years ago, it was an insoluble problem.

The mystery of the curvature of space is of a similar nature. As we have seen the mean density of the universe, the density parameter, or the curvature parameter to astronomers (Ω), lies between 0.1 and 1. Which comes down to saying that space is flat, or almost absolutely flat. In 1979, Robert Dicke and Paul Peebles asked themselves why the universe is so extraordinarily well organized. The two American astrophysicists calculated that, to obtain a value of Ω almost equal to 1 today, its value, one second after the Big Bang, had to be equal to one to the nearest 1 million billionth part!

For some scientists, the

enigmas of the cosmic horizon and curvature are non-existent problems. They maintain that the current universe is simply the result of the initial conditions in the Big Bang. If, they say, Ω had a value of 0 or of 1000, there would be no cosmologists around to measure its value. In the first case, the universe would be empty, and in the second it would have no sooner emerged from the Big Bang that it would have disappeared in an immense symmetrical implosion. The majority of cosmologists, however, consider that these two properties of space-time, the homogeneity and absence of curvature, have the same, fundamental, physical cause.

INFLATIONARY MODELS OF THE UNIVERSE

At the beginning of the 1980s, new models of the Big Bang were proposed, first by the Russian Alexei Starobinski, then by the Americans Alan Guth and Paul Steinhardt, and finally by the Russian Andrei Linde. These researchers discovered from their equations of the primordial universe that the expansion would not have occurred at a steadily increasing rate. The very structure of space-time, around the time 10^{-35} s, would have spontaneously caused an exponential acceleration in the cosmological expansion. This incredibly fast event – called inflation by the scientists – may be very loosely compared with the phase change that occurs in water when, at $0°C$, it goes from the liquid state to a solid. This sudden change in regime undergone by the expansion of the universe did not last for more than an infinitesimal instant: at time 10^{-35} s inflation ceased, and the universe reverted to the slow expansion that we still see today. The cosmologists calculate that in that tiny interval, the size of the universe increased by an incredible amount, between 10^{25} and 10^{100} times! Among their other advantages, these inflationary scenarios allow researchers finally to explain the universe's homogeneity and absence of curvature. In effect, if the primordial universe did undergo such an episode of inflation, the expansion of space occurred infinitely more rapidly than the expansion of the cosmic horizon. To put it another way, our current horizon – which, to have some yardstick we may say is some twelve billion light-years – includes an infinitesimal part of the real universe. All the regions of the

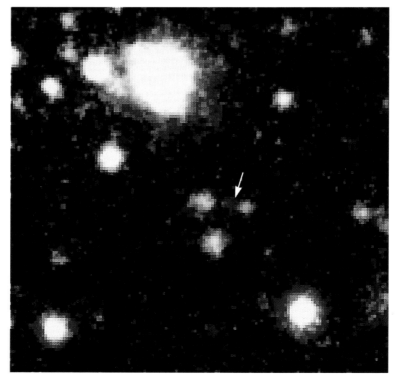

■ DISCOVERING WHEN AND HOW THE FIRST GALAXIES WERE BORN IS ONE OF THE GREATEST PROBLEMS IN COSMOLOGY. TO RESOLVE IT, ASTRONOMERS USE GIANT TELESCOPES, WHICH ARE THE ONLY ONES CAPABLE OF DETECTING THESE DISTANT, FAINT OBJECTS. THIS IS THE GALAXY RD1, AT z = 5.34, AS SEEN BY THE KECK II TELESCOPE.

universe visible today originated from a single volume of the primordial universe, within which they were causally related before inflation. Inflation, by instantaneously multiplying the scale of the universe by several billion, billion, billion, billion times, literally wiped out whatever curvature that it may have had at the time of the Big Bang. The surface of a tennis ball appears strongly curved but if, in an instant, the ball were to be inflated to the size of the Earth, the curvature would no longer be measurable. With this difference: the size of the universe, before and after the inflationary episode, increased by infinitely greater proportions than those of a tennis ball compared with those of the Earth...

Inflationary theories are extremely attractive to many scientists; unlike certain cosmological models, they may be tested by observation. Naturally, their most important prediction concerns the curvature parameter: if the universe did really undergo inflation, then $\Omega = 1$. As we have seen, however, astronomers are not currently able to determine this value. The most recent estimates, in fact, suggest that the curvature parameter is certainly close to 1, but is most likely to lie between 0.1 and 0.3.

It may well be Albert Einstein himself who will get cosmology out of this impasse. When, in 1917, Einstein put forward the very first cosmological model, he introduced a term λ – subsequently called the cosmological constant – into his equations, designed to ensure that the universe would be static, rather than – as the father

of relativity had discovered from his equations – naturally unstable. Subsequently, after Hubble's work showed the expansion of the universe, Einstein regretted having introduced this term λ, which he judged to be without any physical significance, and certainly without any cosmological justification. Curiously, advances in modern physics may perhaps confer a legitimate status on Einstein's cosmological constant. Theorists are, in fact, beginning to wonder if λ is not one of the fundamental cosmological parameters, along with H_0, the Hubble Constant, Q_0, the deceleration parameter, and Ω, the curvature parameter. Contemporary cosmologists do, in fact, see the cosmological constant as a field of energy inherent in space-time itself, which, after a fashion, plays the part of accelerating the expansion. It is extremely tempting for theoreticians to incorporate the cosmological constant into their equations, because this does – at least with certain specific values – offer the considerable advantage of resolving two important cosmological problems. The first of these concerns the divergence between the estimate of the age of the universe from its oldest stars and the estimate from the Hubble Constant. A non-zero value for the cosmological constant would enable them to 'age' the universe by several billion years, and thus reconcile the cosmological age with the age of the stars.

Above all, however, in the relativistic equations the cosmological constant enters into the calculation of the curvature parameter, which is then written ($\Omega + \lambda$). In other words, if the mean density of the universe does not exceed $\Omega = 1$, all we have to do is to add a value $\lambda = 0.7$ to obtain the curvature parameter

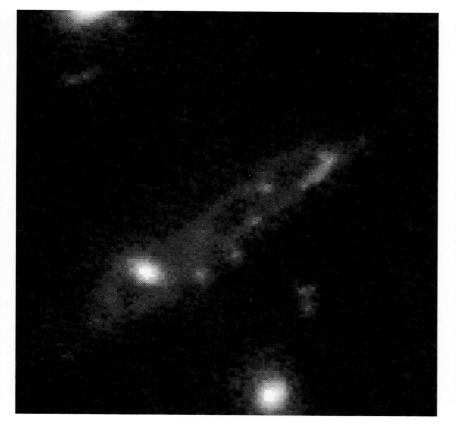
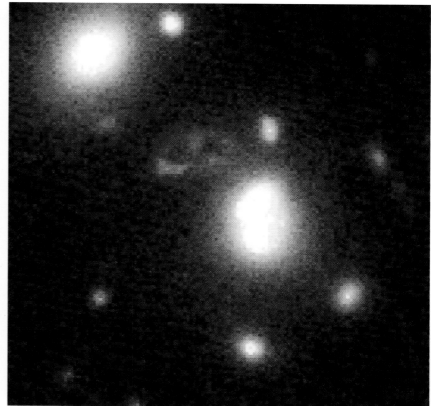

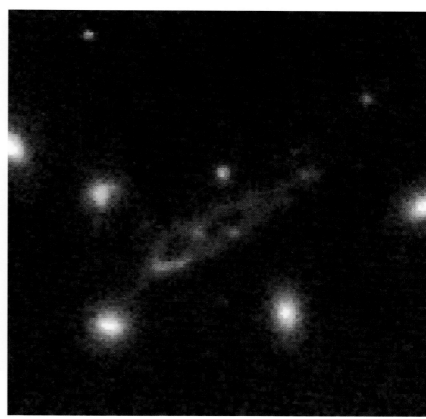

of 1 that is so attractive to those who accept inflationary models.

Like H_0, Q_0, and Ω, λ is a measurable physical quantity. Between 1995 and 1998, two international teams, the Supernova Cosmology Project and the High Z Supernova Team have accurately measured the apparent brightness of several tens of distant supernovae. According to the astronomers, these supernovae are excellent cosmic beacons, because their actual brightness is always the same. The specialists believe, in fact, that measuring the apparent brightness of the supernovae should allow them to measure the parameters Ω and λ, via the relativity equations. They have, in fact, found a non-zero value for the cosmological constant. According to them, $\lambda = 0.7$, and although this is a preliminary result, requiring confirmation, it

■ THE GRAVITATIONAL-LENS IMAGES PRODUCED BY THE CLUSTER CL 0024+1654: FOUR ALMOST IDENTICAL IMAGES — TWO OF WHICH ARE REVERSED, AS IF SEEN IN A MIRROR — OF AN EXTREMELY DISTANT GALAXY, WHOSE REDSHIFT IS STILL UNKNOWN. THESE EXTRAORDINARY IMAGES ARE DEFINITIVE PROOF THAT WE LIVE IN CURVED SPACE-TIME.

is nevertheless encouraging for inflationary models.

No one knows today if the elegant inflationary models do account for the real world. On the other hand, the image that they present of the universe is a fascinating, and almost incredible one, that is actually beyond human comprehension. We have become accustomed, over the course of this last century, as the power of telescopes increased and the limits of the universe seemed to get closer and closer, to being able to take in the whole universe in a single glance. This was, among other aims, the avowed objective of the Hubble Deep Field, which provides us with an image of the universe spanning some 90% of its history. The visible universe is a sort of spatial and temporal sphere, centred on the Earth, about 12 billion years in radius (and as many light-years, if we simplify things by ignoring

the distortion introduced by the expansion), and populated by some 100 billion galaxies. But this sphere is limited by the horizon, which is expanding at the speed of light. If the inflationary models are correct, this bubble – which we have become accustomed to visualizing as the entire universe – is nothing more than a single molecule of water would be, lost in the middle of the Pacific Ocean. Alan Guth believes that, beyond our own horizon, the cosmos carries on extending in all directions. This cosmos is identical to the region of the universe that we can see, it is similarly populated by galaxies and clusters of galaxies and has identical physical properties. The only thing is that it is now inaccessible to science, because our horizon, moving at the speed of light, discloses it very slowly, at the rate of one light-year per year. Theoretical models cannot accurately predict the dimensions of this hidden universe, which emerged from the blindingly fast inflationary phase at the beginning of history, and we are not even close to being able to determine them.

AN INFINITE AND
ETERNAL UNIVERSE

According to Alan Guth, the diameter of this region of the cosmos that resembles our local universe is probably far in excess of 10^{37} light years, which comes down to saying that the cosmos, with physical characteristics like those with which we are familiar, is, at the lowest estimate, one billion, billion, billion times larger than the visible universe. The idea of knowing what happens even farther away is futile. Inflationary theories predict the existence of other domains in the universe, causally disconnected with our own cosmos, and possibly having different physical laws from our own, and these extend – to infinity. But this fascinating hypothesis can probably never be confirmed.

No more, doubtless, than the most extraordinary of cosmological conjectures proposed by the Briton Stephen Hawking, the Americans Edward Tryon and James Hartle, and the

Russian Aleksandr Vilenkin, will ever be confirmed. To these physicists, the Big Bang and inflation are in fact one and the same phenomenon, the spontaneous birth of space-time and the energy that it contains – the entire universe, in fact – from a quantum fluctuation. According to the theory of quantum mechanics, even if it contradicts all logic and common sense, the problem of the chicken and the egg does not arise. Particles can arise out of nothing for no specific reason. These new theories of the Big Bang, built up on the still rather fragile basis of quantum gravity, have thus incorporated this behaviour, which may seem disconcerting or even unacceptable to us, but which it perfectly ordinary in the realm of the infinitely small. Some physicists now accept a seemingly unlikely property of the universe: it may have arisen from nothing.

Recently, Andrei Linde has suggested a further cosmological variation that is even more mind-blowing. In his theory of a 'perpetually self-reproducing inflationary universe', the Russian physicist compares our universe to a Russian doll. According to Linde, our cosmos is nothing more than an infinitesimal bubble lost in an infinite universe, which we should perhaps call the multiverse. The universe would have been born from a quantum fluctuation of another universe, which would itself have arisen from yet another universe... Quantum fluctuations in the space-time of our own universe will also occasionally give birth to other universes, and so on.

Andrei Linde's universe has the astonishing property of removing the problem of the origin of the universe, and of reinstating the perfect cosmological principle. For Linde, in fact, the hypothesis of a beginning is futile, and the Big Bang, which for many years we have associated with the creation of the universe, is nothing more than a single episode in a universe that perpetually regenerates itself, and is thus infinite and eternal. ■

The apparent circling of the stars around the North Celestial Pole, as recorded during a long photographic exposure from the south of the Sahara. The universe still harbours numerous secrets, which astronomers of the third millennium will perhaps be able to solve. When and how were the first galaxies formed? What does the missing mass consist of? What are the age and mean density of the universe? Is the universe infinite and eternal?

THE WORLD'S GREAT OBSERVATORIES

The immense hall is freezing cold, pitch black, and seems to be empty. The metal dome that protects it, as large as a cathedral, is open to the sky and the faint light from the stars enables you to see the outline of the enormous, imposing, and slightly tilted instrument at the centre of the building. A gentle purring is audible, coming from the giant instrument, which gives the false impression of being immobile, but which is slowly turning on its axis, hundredths of a millimetre by hundredths of a millimetre. Higher up, a dull groaning and a dry creak from a metal panel indicate that the dome has also moved.

The long open-work metal tube is pointing towards a spot in the sky that is devoid of stars and lies on the borders of Cetus and Eridanus. There is the slightest variation in the pitch of the purr, a ticking sound, a short click as a cover opens and closes again, and a vague murmuring from the film of pressurized oil, which is all that supports the whole vast mass of metal – the instrument appears to be alive. Stubbornly, the 3.60-m mirror of the Canada-France-Hawaii Telescope remains pointing at the same region of the sky, where a cluster of distant galaxies, Abell 370, is lurking. But no one is present alongside the giant instrument in its dome, which reverberates with a thousand tiny noises.

Because astronomers no longer observe the sky. Or, more correctly, they no longer venture into the rather daunting semi-darkness inside the gigantic domes, and no longer cautiously clamber up to the end of the long, cold tube of the telescope, and finally peer through the eyepiece. Nowadays everything takes place in the control room below the telescope, on a lower floor of the 40-m tower that is dominated by the immense dome. There, in the comfort of a heated, softly lit room, often immersed in the adagio of a Mahler symphony or a long improvisation by the Charlie Mingus' quintet, the telescope's controller sits in front of a keyboard, watching several screens. On this one, atmospheric data – speed and direction of the wind, temperature, pressure, humidity – on the next, the telescope's engineering data, and finally, on the third screen, a portion of the field that the telescope is monitoring, with, right in the centre, the guide star, which dances about and scintillates as a result of atmospheric turbulence. Alongside the controller, an astronomer is absorbed in screeds of numerical data on yet another screen. This is where he will eventually see, after tens of minutes have passed, the image of the cluster of galaxies that he has patiently been recording since the night began.

■ The Canada-France-Hawaii Telescope has a 3.6-m mirror, and is sited at the top of Mauna Kea. It is amongst the telescopes with the highest performance in the world.

From time to time he glances at the guide star that his companion is monitoring. A useless precaution: the telescope is automatic. With an extraordinary degree of precision, it follows the star's apparent motion – and thus the cluster of galaxies at which it is pointing – as the Earth's rotation carries it across the sky. If necessary, however, the telescope's controller can correct the motion of the telescope manually, using a joystick just like that for a video game. Here, however, it is a telescope 15 m long, weighing more than 300 tonnes, that the technician manoeuvres with a light pressure of his thumb.

There are, at the end of this century, less than ten telescopes capable of the performance achieved here, in the shadowy darkness in the dome of the CFH telescope. This instrument, commissioned in 1980 at the top of the extinct volcano of Mauna Kea, on the island of Hawaii, is probably one of the finest ever built.

Yet, by 1996, the CFH was no more than tenth in the list of giant telescopes. Its mirror, just 3.6 m in diameter, gathers half the light of the venerable 200-inch (5-m) on Mount Palomar. But size isn't everything; no instrument more monumental than the Hale Telescope has ever been built, and probably none ever will be. Progress has come elsewhere: rather than conceiving larger and larger and more complex instruments, that the engineers might not actually be able to build, and which no one in any case could afford to finance, astronomers have preferred to opt for better site quality, and an increase in the optical and mechanical precision of their instruments. Above all, however, they have managed to increase the sensitivity of their detectors by an incredible amount.

Nowadays, photographic plates have been replaced to great advantage by electronic detectors: CCDs. These are photosensitive matrices, which convert incident photons into electrical signals. The latter are recorded on magnetic media, which, naturally, means that they may be easily manipulated and analyzed. CCD cameras, with a quantum efficiency of around 80%, are more than ten times as sensitive as the best photographic emulsions. With a CCD detector, for example, the faintest galaxies that may be recorded with the CFH telescope nowadays approach magnitude 28. This gain of five magnitudes, by comparison with the performance attainable in the 1950s, corresponds to the detection of objects that are one hundredth of the brightness!

Between 1947, the year the 200-inch

■ THE MAUNA KEA VOLCANO IS THE HIGHEST POINT OF THE ISLAND OF HAWAII. THIS PICTURE SHOWS THE DOMES, FROM LEFT TO RIGHT, OF THE CFH TELESCOPE, THE NASA 3-M TELESCOPE, THE TWO 10-M KECK TELESCOPES, AND THE 8.3-M SUBARU TELESCOPE.

telescope was commissioned at Mount Palomar, and 1980, when the CFH telescope (the last of that particular line of classic large instruments) began operation, astronomers actually saw the size of their telescopes decrease and their power – at least theoretically – diminish. This was unprecedented in the history of astronomical observation. But by developing the almost ideal astronomical detector in the CCD camera, the engineers also took telescopes extremely close to their theoretical optical limits. By doing so, they rekindled the desire for even larger telescopes. To penetrate even farther into the depths of space and get ever more details of distant objects, astronomers have been forced to design telescopes of even greater diameter.

The 1980s thus saw a true cultural revolution among the astronomical community, which simultaneously abandoned two of the great technological dogmas that it had followed for more than three centuries: the equatorial mounting and rigid mirrors.

THE NEW TELESCOPES

The new-generation telescopes are, simultaneously, less expensive, lighter, more compact, more stable, and far more powerful than their predecessors. First, the mirror. The form

found in a classic instrument such as the CFH is a very rigid disk of Pyrex or glass-ceramic that is some 60 cm thick for a diameter of 3.60 m. It weighs 14 tonnes. The accuracy of its parabolic shape is of the order of 10 millionths of a millimetre, which is attained by the quality of the polishing and then maintained by the strength and rigidity of the glass. By contrast, the largest mirrors currently in the process of being polished are in the form of a thin meniscus, 8.20 m in diameter and only some 20 cm thick. With a surface five times that of the CFH mirror, their weight is under 20 tonnes. Seven, thin, practically identical, 8.20-m mirrors are currently being manufactured. As an inevitable corollary of their extreme thinness and of their light weight, they are flexible, but their fragility is no longer a handicap for the opticians. The latter support the flexible mirror by means of hundreds of microscopically adjustable hydraulic pistons that are controlled by a fast computer. The shape of the mirror is checked and adjusted once a minute, so that, whatever the observing conditions, it maintains its perfect parabolic shape and gives perfect images of the sky. It is extremely difficult to visualize the extraordinary degree of accuracy that the opticians manage to achieve with these large optical components. If we compare the surface of the CFH mirror with that of a lake 360 km in

185

■ IT IS AT LA SILLA, AT AN ALTITUDE OF 2400 M, IN THE CHILEAN ANDES, THAT THE LARGEST CONCENTRATION OF TELESCOPES ON EARTH MAY BE FOUND. THE SITE WAS FOUNDED BY THE EUROPEAN SOUTHERN OBSERVATORY, AND CURRENTLY HAS 13 ASTRONOMICAL INSTRUMENTS.

diameter, the defects on the surface of the glass would correspond, on the lake's scale, to waves in the water that are less than 1 mm high.

The design of the skeleton telescope tubes has also seen a radical change. All modern telescopes are of the altazimuth type, where there are two axes, one vertical, one horizontal. This prevents the mirrors from being subject to the flexure and misalignment that are caused by the slow tilting found on equatorial mounts as they rotate about their main axis. For the engineers, there is an immeasurable gain in size, weight, and price. Yet it was impossible to envisage any such technological solution a few decades ago. With an equatorial mounting, following the stars was both natural and elegant, because its axis simply turned in the opposite direction to the rotation of the Earth. In an altazimuth, this has been replaced by three simultaneous rotations. The control of these three movements, whose speed varies all the time, requires the use of a computer.

At the end of the day, all the giant telescopes at the beginning of the 21st century will have a smaller mass and size than the 200-inch (5-m) at Mount Palomar, yet have mirrors that capture 10 to 20 times as much light.

Naturally, the observation of objects that are 500 million times fainter than the faintest star visible to the naked eye imposes some drastic constraints: during the exposure, which may last tens of minutes or even hours, the sky needs to be crystal clear, and perfectly calm.

It is not, in fact, just its technical capability that distinguishes the CFH telescope from the dozen other instruments with equivalent optics that have been built more or less all over the world. Sited at an altitude of over 4200 m on the highest point of the Hawaiian Islands, the CFH has an exceptionally pure sky for observing. In this extremely dry, high-altitude desert, isolated from the rest of the world in the middle of the Pacific, this Franco-Canadian instrument operates in a rarefied atmosphere that is about half the density of that at sea level. Water vapour, which absorbs some of the infrared and ultraviolet radiation, and aerosols, which affect the sky's clarity, are trapped in the clouds and fogs that continuously form lower down the slopes of the extinct volcano. At night, the sea of clouds that forms beneath the summit of the giant shield volcano damps down atmospheric turbulence and blocks the light from the shoreline villages, which might otherwise interfere with the observation of very faint objects.

The summit of Mauna Kea is the finest site that astronomers

■ SOME OF THE TELESCOPES AT LA SILLA ARE DEVOTED TO SPECIFIC RESEARCH PROJECTS. THE TWO 1-M TELESCOPES OF THE DENIS AND EROS 2 PROJECTS ARE USED, RESPECTIVELY, FOR INFRARED MAPPING OF THE SOUTHERN SKY, AND THE SEARCH FOR THE MISSING MASS.

have found in the Northern Hemisphere. Over the course of time, since its discovery at the beginning of the 1960s, larger and larger, and more and more powerful instruments have been installed there. The 3.60-m diameter CFH telescope of the Canada-France-Hawaii Institute has been joined by the two Californian instruments of 10 m in diameter, currently the largest in the world.

These twin Keck telescopes are in the forefront of technology, conclusively breaking the classical canons of optics. Optical workers are currently unable to melt, cast, machine, and finally, accurately polish monolithic astronomical mirrors more than 8.30 m in diameter. To construct instruments of even larger diameter, however, the American scientists had the revolutionary idea of creating a hexagonal mirror 10 m across, using a mosaic of 36 small hexagonal mirrors, each 1.80 m across, butted up against one another. The hyperbolic shape of this segmented mirror, 10 m in diameter, yet only 7.5 cm in thickness, is monitored twice a second by 168 sensors to an accuracy of nanometres, and subsequently adjusted by 108 jacks.

Before the end of the millennium, two other new instruments, the Japanese Subaru telescope, 8.30 m in diameter, and Gemini 1, an international, 8.20-m telescope, will be operational at the Hawaiian observatory.

In a few years' time, there will be only a few major, international astronomical centres. Modern telescopes demand skies that are of an all-too-rare purity and stability. In the Northern Hemisphere, Mauna Kea has garnered most of the giant telescopes. It is by far the best-equipped observatory in the world. Some important sites also exist in Arizona, like the Kitt Peak Observatory, and the newly founded observatory on Mount Graham, at an altitude of 3300 m, or Mount Hopkins, where a 6.50-m telescope was commissioned in 1998. As far as Europe is concerned, atmospheric conditions almost as good as those at Mauna Kea have been found in the Canary Islands, where, at an altitude of 2400 m, the Tenerife and La Palma observatories have been founded.

The Southern Hemisphere has only recently seen the development of major observatories, despite the fact that it has some of the most interesting objects in the sky that are without parallel in the north, such as the two Magellanic Clouds, extremely close galaxies, or the globular clusters Omega Centauri and 47 Tucanae, which are exceptionally rich in stars. With the notable exception of the Anglo-Australian Observatory, at Siding Spring, sited in a forest of eucalyptus at an altitude of just 1150 m, all the major observatories in the Southern Hemisphere are in the same region: the Chilean Andes. The United States, in collaboration with various South-American countries, and Europe in collaboration with Chile have set up a total of five major observatories at altitudes over 2000 m, all along the Pacific coast, between the cities of La Serena and Antofagasta. The three sites Cerro Tololo, Las Campanas, and La Silla have some twenty telescopes between them, and in the last twenty years have been visited by thousands of astronomers. It is impossible to keep track of the number of discoveries that have been made from these sites, but the most outstanding was the extraordinary supernova in the Large Magellanic Cloud in February 1987, which could be followed only from the Southern Hemisphere.

Two new sites are currently being developed for new giant telescopes. At the summit of Cerro Pachon, an international team is building the second of the 8.20-m telescopes for Project Gemini, as Gemini 1 is being commissioned in Hawaii. These twin telescopes, designed to provide complete coverage of both the northern and southern skies, will begin operation early in the 21st century.

THE GREAT OBSERVATORIES

The most ambitious project for the beginning of the third millennium is undoubtedly the European one. The European Southern Observatory's Very Large Telescope (VLT) is being constructed on the summit of Cerro Paranal, an isolated mountain, 2500 m high, in the middle of the Atacama Desert. It is currently the only astronomical site that can rival Mauna Kea. Permanently occupied since 1983 by a team of meteorologists, it has revealed exceptional qualities: an atmosphere that is as calm and transparent as that of Mauna Kea, and an extraordinarily high number of nights favourable for observation. The sky is clear at Cerro Paranal on 340 nights in the year! The VLT is a set of four 8-m telescopes, installed on the 1-hectare summit platform of Cerro Paranal, which had to be reduced in height by some thirty metres to accommodate it. The four telescopes may be used as four independent instruments, or simul-taneously observe the same astronomical object. In the latter case, the sensitivity of the VLT will be the same as that of a telescope 16 m in diameter. This gigantic astronomical instrument will be progressively commissioned between 1998 and 2001.

Mauna Kea, in the Northern

■ THE ATACAMA DESERT SEEN FROM CERRO PARANAL. IN THE BACKGROUND LIES A VERY PROMISING FUTURE SITE FOR ASTRONOMY: CERRO ARMAZONES, A MOUNTAIN 3000 M HIGH.

Hemisphere, and Cerro Paranal, in the Southern, promise to be the twin poles of observational astronomy in the decades to come. Despite the large number of clear nights that they experience, these observatories will never offer astronomers ideal conditions. Terrestrial telescopes, whatever the diameter of their mirrors, are limited because they are affected by atmospheric turbulence, especially when they are recording the faint light from celestial objects for hours on end. When a ray of light crosses the last 20 000 metres of the Earth's atmosphere it encounters masses of air at different temperatures and pressures, constantly mixed by turbulence, or broken up by the wind. For every slight change in the atmosphere's refractive index, the light is slightly deviated and the image is degraded, like a reflection in a pool of water ruffled by the wind. To the naked eye, the turbulence appears in the form of slight scintillation of the stars. But when it is amplified 100 or 1000 times in a telescope, the rather attractive twinkling is transformed into a boiling motion from which it is impossible to extract the slightest information. All the difficulty of astronomical observation lies here: the larger the telescopes, the more powerful they are; but the more powerful, the more sensitive they are to atmospheric turbulence.

THE SPACE TELESCOPE

The most radical method is, of course, to send a telescope up beyond the atmosphere, into the vacuum of space. This was how the Hubble Space Telescope project was born. For more than twenty years this instrument has carried all of the astronomers' dreams of exploration. It was conceived by American scientists in the 1960s, and its construction was agreed by NASA and ESA in 1977. It would be a classical telescope, fitted with a 2.4-m mirror, and carrying four instruments, including two cameras, one American, and one European. Although the size of the HST's optics may seem modest, its observing site, in Earth orbit at an altitude of 600 km, ought to give it a completely clean, black sky, and one that would, of course, be utterly free from turbulence. The complete instrument, 14 m long, and weighing 12 tonnes, was not finished in time for its original commissioning date, which was initially set for 1983. In 1986, the explosion during the launch of the Space Shuttle *Challenger* that caused the death of seven astronauts, stopped the American manned space programme in its tracks, and astronomers had to wait patiently for four more years, awaiting

■ THE NTT (NEW TECHNOLOGY TELESCOPE) WAS COMMISSIONED IN 1990 AT THE LA SILLA OBSERVATORY. ULTRA-LIGHT, AND COMPACT, THIS INSTRUMENT, WITH A 3.5-M MIRROR, FORESHADOWS THE FOUR 8-M TELESCOPES OF THE EUROPEAN VLT (VERY LARGE TELESCOPE) PROJECT, WHICH WILL BE INSTALLED AT CERRO PARANAL EARLY IN THE 21ST CENTURY.

189

the resumption of space flights. Finally, in April 1990, the long-awaited telescope was launched on board the Space Shuttle *Discovery*. What a disappointment! The astronomers were appalled to find that the parabolic shape of the mirror had been given too steep a curve during polishing, thanks to faulty optical tests. The Hubble Space Telescope, the most expensive scientific instrument ever built – its total cost exceeded 3 billion dollars – was short-sighted. The telescope was finally repaired in orbit, in December 1993, during a visit by the seven astronauts on board the Shuttle *Endeavour*.

Since then, the Space Telescope has fulfilled to perfection the programme for which it was conceived, that of scrutinizing the cosmic horizons with unprecedented accuracy. Its performance far surpasses the best ground-based telescopes. In an exposure of four hours, its cameras are able to record objects at magnitude 27, and, in 20 hours, it can reach magnitude 29. Finally, and above all, this exceptional sensitivity is accompanied by a resolution that has never been attained by ground-based instruments. Currently, the HST's light-grasp, in comparison with the CFH telescope, for example, is more than one magnitude greater. This enables it to discover galaxies that are three times as far away, for example, or see stars, clusters, nebulae or galaxies clearly, where a ground-based telescope would see little more than an indistinct blur.

Despite its unique capabilities, the HST is not able to carry through all the observational programmes that the scientific community would like to achieve. The only optical telescope in orbit, it is used indiscriminately to study everything in our own galaxy, stars in nearby galaxies, or the most exotic and distant objects, such as clusters of galaxies, quasars or gravitational lenses. This is undoubtedly far too many targets for a single instrument, even if it were the best in the world. The field covered by the HST is, in fact, minute. It is a small square of less than 3" × 3" on the celestial sphere. If astronomers decided to cover the entire sky with the HST, giving, let us say, eight hours' exposure for each field, it would take them 50 000 years of continuous observation!

Finally, although no one questions the absolutely incomparable quality of the images obtained by Hubble, the specialists are aware that its relatively modestly sized mirror does not allow it to make the most demanding scientific observations, because the latter require a large amount of light. In other words, astronomers may be partial to fine astronomical photographs but they are particularly interested in analyzing the objects under consideration. Only spectroscopy provides the chemical composition of objects, their speed of rotation, and their distance, but because of its dispersion – the way in which the light is spread out – a spectrum is extremely greedy for photons. As a proof of this, no spectrograph is currently capable of recording the light from the faintest stars and galaxies accessible with the CFH telescope or the HST. Astronomers are certainly able to photograph objects of magnitude 28, 29, or even 30, but they are unable to determine their physical nature. The situation is extremely frustrating, and has provoked the construction of gigantic photon collectors like the Keck, Gemini, Subaru, and VLT telescopes. The Keck telescopes, for example, have a mirror surface of 80 m^2, which captures 3 million times as many photons per second as the human eye! When it comes to obtaining a spectrum, the capacity of a telescope for accumulating as much light as quickly as possible is the sole governing factor, so astronomers often prefer to use large ground-based telescopes in preference to the HST. For spectroscopy, the CFH alone is twice as fast as the Hubble Space Telescope. As for the Keck telescopes, their 10-m mirrors enable them to obtain a spectrum in less than an hour that the HST would have difficulty in recording with a sixteen-hour exposure. The demand for time on the only optical space telescope is so great among the scientific community, however, that a team is practically never awarded such an amount of time for observation. In fact, it is on Earth that the future of astronomy will be played out: at the beginning of the 21st century, observers will have twenty telescopes 2.50–5 m in diameter, and fifteen 5–10 m. This will be enough to capture several hundred times as many photons as the single HST, which suddenly appears rather lonely in its distant orbit.

Being forced to abandon any hope of installing their observatories and precious instruments in space, where they would have benefited from ideal observing conditions and improved the optical performance of the latter literally ten- or even a hundred-fold, researchers decided, at the beginning of the 1980s to go to the root of the trouble, and eliminate the deleterious effects of atmospheric turbulence.

The technique that was developed in various European and American laboratories, uses what are called 'adaptive' optics. These are systems that employ an image-analyzer, an ultrafast computer, and a small, flexible mirror, with a surface that may be deformed, that is mounted on a network of pistons. An adaptive-optics system operates in a very simple and clever manner. The telescope observes a star and records an ultrafast succession of images. A computer linked to the camera analyzes the image, which

■ LAUNCHED BY THE SPACE SHUTTLE *DISCOVERY* IN 1990, THE HUBBLE SPACE TELESCOPE HAS BEEN SERVICED TWICE BY ASTRONAUTS, IN 1993 AND 1997. THE NEXT VISITS WILL OCCUR IN 1999 AND 2002.

scintillates, changes in shape, and even moves about as a result of atmospheric changes. The computer compares this image with the theoretically perfect point-like image that the telescope would capture in the absence of turbulence. It automatically orders the deformable mirror, which is situated in the beam of light coming from the star, to correct the image, until it approaches that of a theoretical star. In a few seconds, everything is sorted out: the mirror is deformed under computer control, and the star resumes the appearance it had before it was degraded by the atmosphere. The analyzer checks the appearance of the

■ FROM ITS 600-KM ALTITUDE ORBIT, THE HUBBLE SPACE TELESCOPE IS ABLE TO OBSERVE A SKY THAT IS TRANSPARENT AND FREE FROM TURBULENCE. THIS INSTRUMENT, WHOSE MIRROR IS JUST 2.40 M IN DIAMETER IS CAPABLE OF DETECTING 30TH-MAGNITUDE GALAXIES.

new image, transmits it to the computer, and so on.

Various adaptive-optics systems are beginning to be used, in particular the Pueo instrument on the CFH telescope, and Adonis on the 3.60-m telescope at La Silla in Chile. These two telescopes have obtained images of double stars, of nebulae, and even of galaxies that are comparable with those from the HST. Systems with deformable mirrors are, however, difficult to commission and are operational in the observatories for only a few tens of nights every year. It will probably take scientists another ten years of development to ensure that these detectors are as

■ THE KECK I TELESCOPE AND ITS TWIN, KECK II, ARE THE MOST POWERFUL ASTRONOMICAL INSTRUMENTS OF THE AGE. INSTALLED AT THE MAUNA KEA OBSERVATORY IN 1991 AND 1996, RESPECTIVELY, BY CALIFORNIAN ASTRONOMERS, THEY HAVE HEXAGONAL MIRRORS 10 M IN DIAMETER. IT IS THE KECK TELESCOPES, WHICH ARE PRIMARILY USED FOR SPECTROSCOPIC WORK, THAT HAVE OBSERVED THE MOST DISTANT GALAXIES IN THE UNIVERSE.

reliable and practical as a simple CCD camera. By then the Space Telescope will have retired, and astronomers will be making full use of adaptive optics with their new, giant telescopes, finally free of the unpredictable effects of the atmosphere.

RADIOTELESCOPES

Light – that is the radiation visible to the human eye – is not the only carrier of information in the universe. The light that astronomers capture with 'optical' telescopes represents a tiny fraction of the immense range of electromagnetic radiation emitted by astronomical objects. After 500 million years of vertebrate evolution, the human race has been endowed with an image receptor whose sensitivity is perfectly adapted to the range of radiation emitted by the star that shines down on our planet. As a result, our eyes are able to see radiation with wavelengths between 400 and 700 nm (nanometres) – light 'waves' around 0.5 thousandths of a millimetre – the region in which the Sun emits the most energy. The brain translates these into colours, violet and blue for the shortest wavelengths and red for the longest. Although our vision may be limited, the Sun and other astronomical objects produce a whole range of radiation that is far more significant, despite being of precisely the same nature as visible light. The more energetic the radiation, the greater the frequency, and the shorter the wavelength. Beyond the visible region there is the ultraviolet region, followed by X-rays, and finally gamma-rays, produced in the cores of stars or other energetic processes. That's the high energies. On the other side of the visible region, there is the infrared. Although infrared rays from the Sun pleasantly warm our skin in the summer, it is the ultraviolet radiation that causes it to tan, or even inflicts dangerous sunburn. Finally, beyond the infrared, we come to the limitless region of radio waves, where radio-astronomers operate.

Obtaining clear images in the radio region is difficult. The resolving power of a telescope is proportional to the diameter of its mirror and the wavelength of observation. Optical telescopes observe light whose wavelength is tiny: 0.5 thousandths of a mm on average. The radio wavelengths that interest astronomers, by contrast, stretch from 1 cm to 1 m. In concrete terms, that means that, in order to see the sky as accurately at radio wavelengths as in the optical region, astronomers need radiotelescopes with parabolic aerials – mirrors consisting of a fine metal mesh that reflects and concentrates the radio waves – that are of an utterly impractical

size. How, in fact, can one construct an instrument several kilometres in diameter?

of their mirrors remains inadequate.

Astronomers managed, however, to get round this apparently insurmountable obstacle. To start with, radiotelescopes have effectively become gigantic machines, some of the largest ever built. Designed on similar principles to optical telescopes, the radiotelescopes at Parkes, in Australia, at Jodrell Bank, in England, and at Effelsberg, in Germany, have metallic mirrors 64 m, 76 m, and 100 m in diameter, respectively. These aerials – the one at Effelsberg weighs 3200 tonnes – move on two axes, and may be turned in any direction on the sky in just a few minutes. The largest astronomical mirror on the planet is even more impressive: 300 m in diameter. This immense metal mesh has been stretched over a natural depression at Arecibo, in Puerto Rico. The metal mirror is fixed, and it is the radio receiver, hovering between the sky and the ground at the focus of the aerial, in a metal framework held up by a system of cables and three immense concrete towers, that slowly moves to follow the apparent motion of the radio sources across the sky. Despite their great sensitivity, capable of detecting radio signals emitted by objects that are more than 10 billion light-years away, these Cyclopean instruments are still unable to get a clear view of the universe. Despite their enormous size, the diameter

INTERFEROMETERS

Scientists soon found the radical solution; it consisted of synthesizing a giant, virtual telescope, using a set of smaller, but definitely real, telescopes. All the astronomers needed to do was to consider each of their aerials as a tiny element in a large virtual reflector, and to reconstruct the image that this imaginary mirror would give. The larger the number of small individual elements, the better the fit between their distribution and the extent of the virtual reflector, and the more accurate the images or spectra that could be obtained.

These arrays of aerials are known as interferometers. The best-known of these was installed in New Mexico in 1980. The VLA (Very Large Array) records and accurately times the arrival of the radio signal from an astronomical object at the focus of each of its 27 aerials, each 25 m in diameter. A powerful computer then recombines all the data that were recorded separately and reconstitutes the image of the object, as if it had been observed by a telescope with a diameter equivalent to the maximum distance between the aerials. Depending on the wavelength being observed and the resolution required by the researchers, the diameter of the VLA is variable, because the 27

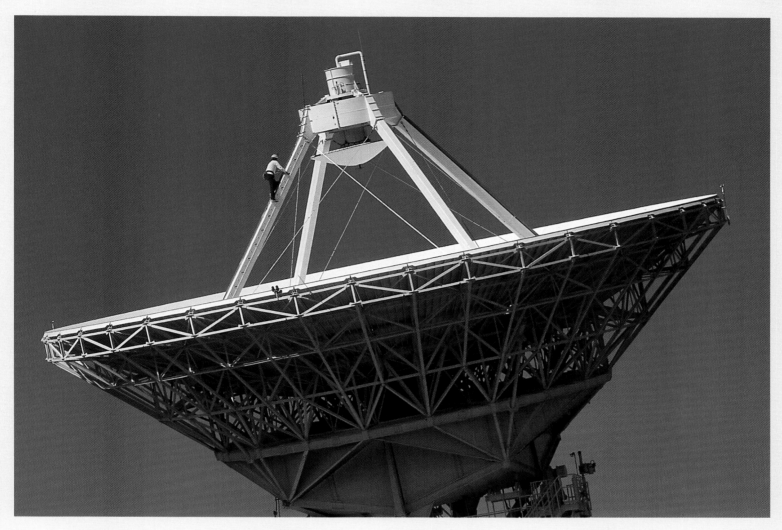

ONE OF THE TEN 25-M AERIALS OF THE VLBA (VERY LONG BASELINE ARRAY). MEASURING 8000 KM ACROSS, THIS INTERFEROMETER, WHICH IS ONE OF THE MOST POWERFUL IN THE WORLD, EXTENDS RIGHT ACROSS THE USA, FROM THE VIRGIN ISLANDS TO HAWAII.

aerials may be moved along railway tracks. The minimum size of the array is 600 m, and the maximum diameter of this New-Mexico giant is 27 km. With such a diameter, the radio portraits of galaxies, in which it has specialized, have the same accuracy as that given by the large optical telescopes. On the other hand, the sensitivity of the VLA, i.e. its capacity to detect very faint sources, is not what would be obtained with an aerial that did actually measure 27 km across. From this point of view, the VLA should really be compared with a photographic lens that is stopped down to an exaggerated degree. The equivalent surface of the 27 aerials in the array is the same as that of a single 130-m radio aerial, more or less equivalent to the actual Effelsberg radiotelescope. The two types of instrument are complementary, however, with interferometers being used for tiny objects – especially the nuclei of galaxies – and the large dishes for wide-field or spectroscopic observations that require the maximum sensitivity.

Despite its size – drivers on Route 60 in New Mexico sometimes drive from side to side of the array of aerials without even noticing the tiny, distant dishes, distorted and wavering with the desert mirage, and apparently hanging above the horizon – the VLA is itself a small element in a far larger interferometer, the VLBA (Very Long Baseline Array). Commissioned in 1994, the largest radio-interferometer on the planet befits the scale of the New World. It measures 8000 km from east to west, and nearly 4000 km from north to south. In the east the VLBA's first aerial points towards the sky from a beach in the Virgin Islands, in the Caribbean. In the west, the last metal dish hides its delicate tracery in a freezing cold, isolated valley, not far from the summit of Mauna Kea, the volcano on Hawaii. Between these two extremes, the radio-interferometer spins its web across the whole of the North-American continent, from New Hampshire to Texas, from Iowa to New Mexico, and from California to Arizona. Although the sensitivity of the ten aerials – to which those of the VLA are often added – does not exceed that of other great international instruments, the VLBA's resolution is unique and duly stupefying: between about 0.03" and 0.0005", depending of the wavelength being observed. With comparable visual acuity, it would be possible for someone in Washington, D.C., to read a book in San Francisco! On the Moon, to take a more astronomical example, 0.0005" represents details less than 1 m across – that's one hundred times better than the Hubble Space Telescope!

Astronomers dream of being able to apply this interferometric

THE INTERFEROMETER ON THE PLATEAU DE BURE, IN THE ALPS, OBSERVES THE MILLIMETRE RADIATION EMITTED BY HIGH-REDSHIFT GALAXIES. IT IS IN THE MILLIMETRE REGION THAT IT IS HOPED TO DISCOVER THE VERY FIRST GENERATION OF GALAXIES.

method – which has allowed the VLBA to penetrate right to the heart of galaxies for the first time, and finally prove the existence of black holes within them – to other regions of the electromagnetic spectrum. Today, radio waves between about 1 mm and 1 m may be studied in this way, but interferometric observation at shorter wavelengths represents a technological hurdle that is only just being overcome, after three decades of research, and the pioneering work by Antoine Labeyrie in France. In the visible and infrared regions, the optical and mechanical precision required of interferometers is more than one thousand times that for radio-interferometers.

Although radio, infrared and visible radiation does reach astronomers' instruments, the same cannot be said for high-energy radiation, which is almost completely blocked by the Earth's atmosphere. This is actually lucky for us, because exposure of the Earth's surface to ultraviolet radiation, X-rays and gamma-rays would have prevented any form of life from arising and developing. This invisible 'light' is emitted by very hot astronomical objects, such as supergiant stars, for example, or black holes. Ever since the 1970s, scientists have tried to gain access to this high-energy radiation and have installed highly complex telescopes on board balloon-sondes, rockets, and satellites. It is extremely difficult to work with these ultra-short-wavelength photons, and traditional optical techniques are hard to apply to the task of capturing particles that are 1000 times as energetic as a simple photon of visible light. Nevertheless, the various satellite observatories – Einstein, Compton, Granat, etc. – most of which give only a very indistinct view of the sky – have given rise to staggering scientific advances by revealing physical phenomena, such as the gamma-ray bursters, that were quite unexpected, and undetectable with ordinary methods.

Nowadays, all the windows on the electromagnetic spectrum, from nanometre to decametre wavelengths, have been opened wide, with the exception of a few that might be said to be ajar. When astronomers are confronted by phenomena that are particularly difficult to understand, they open them all at the same time. It is possible to see, for several hours at a time, the optical and infrared telescopes at the observatory on Mauna Kea, on Hawaii, the VLA in New Mexico, and the Hubble and Compton satellite observatories, all pointing in the same direction, all capturing very different signals, emitted by hotter and hotter, and deeper and deeper layers of gas that hide the fantastic, mysterious heart of some distant quasar. ■

■   NOWADAYS, SCIENTISTS HAVE OPENED UP THE WHOLE RANGE OF THE
ELECTROMAGNETIC SPECTRUM, FROM GAMMA-RAYS TO RADIO WAVES. SHOWN HERE IS
THE 64-M AERIAL AT PARKES, IN AUSTRALIA, WITH WHICH ASTRONOMERS HAVE
DISCOVERED THOUSANDS OF QUASARS.

THE NEXT SPACE TELESCOPE

Since the end of the 1990s, American and European astronomers have been thinking about the successor to the Hubble Space Telescope. This future instrument will be one of the corner-stones of NASA's new research programme, which is dedicated to the origins of planets, of stars, and of galaxies. Like its famous predecessor, it is primarily the cosmological aspect of this ambitious project that will occupy the HST's successor from 2006 onwards. Why is NASA already thinking about a replacement for the Hubble Space Telescope when the latter is delighting researchers with the quality of the observations that it is producing? It is simply that, well before the end of its mission, which is expected to be in 2005, the HST will have obtained the spectacular results anticipated from it, and also, at the same time, shown its limitations. Although all branches of astrophysics are involved in the results from the HST, its particular contribution was expected to be in the field of cosmology. The name chosen for the space telescope reflects the hope that researchers placed in it. Edwin Hubble: the man who revealed the nature of galaxies, and following closely on from that, who discovered the expansion of the universe. Obtaining an accurate measurement of the constant named after him and, at the same time, determining the age of the universe, were the prime objectives of the Space Telescope. In parallel with this, there was the goal of observing the evolution of galaxies far away in space, and thus far back in the universe's past.

This ambitious cosmological programme has undoubtedly had its successes, even though it has proved to be more difficult to carry out than expected, as researchers discovered that the universe was more subtle and devious than they had previously expected. Although the Hubble Space Telescope successfully opened up new astronomical fields that were not even suspected of existing at such distances, it has shown that it is not capable of expanding the limits of exploration any farther. Admittedly, a second servicing mission at the beginning of 1997 did install two new instruments: the STIS spectrograph and the NICMOS (Near Infrared Camera and Multi-Object Spectrometer) instrument. The latter, foreshadowing the NGST (Next Generation Space Telescope), gave the HST infrared vision for the first time, at wavelengths between 0.8 and 2.5 mm (microns). In 1999, the third servicing mission should be able to install another new instrument, NASA's Advanced Camera. The performance of this is expected to be spectacular. It has a larger field than the current WFPC 2 camera, and it will give the HST twice the resolution and greater sensitivity. Finally, European scientists at ESA are designing a new-generation instrument for the Hubble Space Telescope, which will again be visited by the Space Shuttle in 2002. So the HST should be able to continue working at least until it is replaced by the NGST, expected in 2006.

Despite all these improvements, the Space Telescope will remain restricted. Although astronomers praise its good points, they also readily admit to its faults. Its 2.4-m mirror is too small. In addition, launched from the Space Shuttle at an altitude of just 600 km, it is hampered more than half of the time by Earthlight, moonlight, or by the Sun, which prevent it from observing.

Finally, and primarily, the HST is unsuited for accurate observation of the distant prospects that it discovered and, perforce, of the *terrae incognitae* that lie even farther away, back in the universe's past. In fact, the light from galaxies lying more than 10 billion light-years away is shifted, through the expansion of the universe, into the infrared, a region in which the HST is almost blind, because its favoured range extends from the ultraviolet to the visible.

At the beginning of 1996, the 'HST and Beyond' committee, chaired by Alan Dressler, therefore recommended the construction of the NGST, a new space telescope, far more powerful than the Hubble Space Telescope, and sensitive to infrared radiation. The instrument, whose primary goal would be the study of the formation of the earliest stars and galaxies, should be able to determine the various cosmological parameters: the Hubble Constant, of course, but also the deceleration parameter, the density parameter, and finally, the cosmological constant. Which should, in principle, answer the major questions in modern-day cosmology: What is the age of the universe? What is its geometry? What is its future? Naturally, astronomy is not just cosmology, and like the HST today, the NGST would be able to observe all sorts of objects, from nearby stars to quasars, via embryo stars in galactic nebulae, and the nearest extrasolar planets.

As soon as the recommendations of the Dressler report were known, in 1997, the Space Telescope Science Institute (STScI) published a very detailed preliminary proposal, which fore-shadows the probable design of the NGST. Given the reduction in federal budgets, NASA envisages the overall cost to be no more than 1 billion dollars, including launch and support costs, with an estimated lifetime of five to ten years. By comparison, by the

time it retires, the HST will have cost, in total, more than 6 billion dollars. The instrument, which will be launched on a classic rocket of the Titan, Atlas, or Ariane 5 type, will be placed into an orbit where it will be able to observe the sky for the full 24 hours, from the L2 Lagrangian point. This solar orbit, which it will complete, like the Earth, in exactly one year, will keep it on the Sun-Earth line, at about 2 million kilometres from the planet. There, the NGST will permanently turn its back on the Sun, Earth, and Moon, which will all lie in the same area of sky. It will observe the universe 24 hours out of 24 in the opposite direction to the over-bright trio, and the whole of the celestial sphere will be swept over the course of the year, as it is from Earth. There are twin advantages of the L2 Lagrangian point: on the one hand, the sky is much darker from there than it is from the HST's low Earth orbit; and on the other hand, being constantly protected from the Sun, the telescope will be naturally cooled to a very low temperature, which is ideal for observing at infrared wavelengths. The NGST should be sensitive between 0.5 mm and 30 mm (i.e. from the visible to the infrared) and should obtain images as accurate as those of the HST in the visible, and with a comparable field. Apart from its mirror, which out of necessity is segmented, like those of the Keck telescopes, so that it will fit into the nosecone of a rocket, everything in the NGST is simplified. Protection of the instrument is reduced to its simplest form. A two-part sunshield consisting of multiple layers of tensioned, coated Mylar film, will shade the mirror, which will be exposed directly to the vacuum of space. The total mass of the NGST will be less than 2.8 tonnes, whereas that of the HST is 11.6 tonnes. One sixth of the price, one quarter of the mass, the Hubble Space Telescope's successor will nevertheless be far more powerful. Capable of detecting stars and galaxies down to magnitude 32, it should, in principle, be able to follow galaxies like ours – if they existed then - back in time to redshifts (z) of 10 or 15. Another primary objective for this telescope is to observe supernovae in this same range of redshifts, to use their behaviour to determine the curvature of the universe. The first supernovae may have exploded a few hundred million years after the Big Bang, at $z = 10$, 20, or 30, and, according to certain cosmological models, they should be observable at wavelengths of around 4 to 10 microns.

The NGST, with average exposure times of around fifty hours, should be able to detect to the very first glimmers of light from the depths of time. Statistically, astronomers expect to observe one supernova per year, in a field 1' across, at redshifts greater than 5. Over the whole celestial sphere, that represents more than 100 million supernovae per year. Finally, just as the HST has done down to magnitude 30 in the visible region, the NGST will obtain long-exposure photographs of the distant universe through various filters. This will provide 'sections' by space and time at successive redshifts. Again, the aim is to understand the architecture of the universe.

The scientists and engineers will need to draw up the definitive specification for the project very quickly, because their time is limited. Ideally, the NGST should enter service when use of the HST comes to an end, at the earliest in 2005, or, as the Dressler Committee hopes, a few years later. All that remains is to find a name for the new space telescope. That's a difficult task: despite the spectacular progress in observational cosmology, it seems that our times have yet to find their own Edwin Hubble.

THE HERTZSPRUNG-

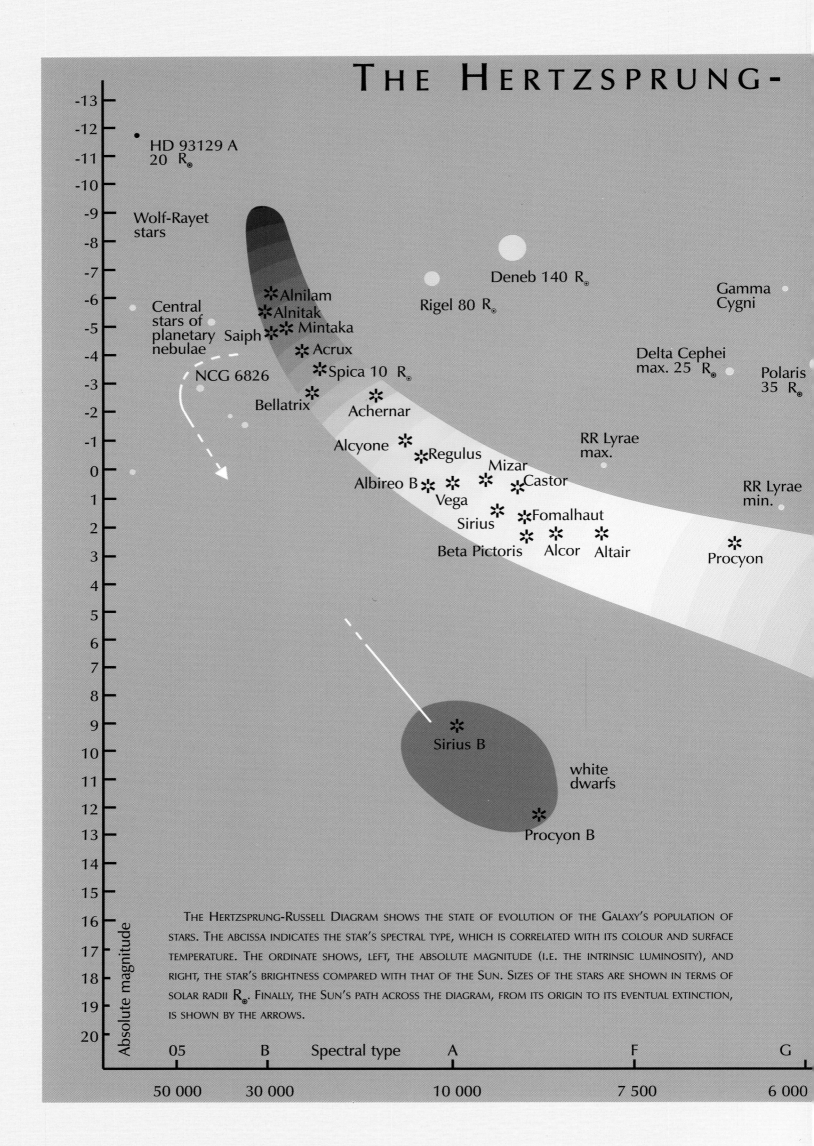

HD 93129 A
20 R⊙

Wolf-Rayet
stars

Central
stars of
planetary
nebulae

Saiph

NCG 6826

Bellatrix

Alnilam
Alnitak
Mintaka
Acrux
Spica 10 R⊙
Achernar

Alcyone
Regulus
Mizar
Castor
Albireo B
Vega
Sirius
Fomalhaut
Beta Pictoris
Alcor
Altair

Deneb 140 R⊙
Rigel 80 R⊙
Gamma
Cygni

Delta Cephei
max. 25 R⊙
Polaris
35 R⊙

RR Lyrae
max.

RR Lyrae
min.

Procyon

Sirius B

white
dwarfs

Procyon B

Absolute magnitude

THE HERTZSPRUNG-RUSSELL DIAGRAM SHOWS THE STATE OF EVOLUTION OF THE GALAXY'S POPULATION OF STARS. THE ABCISSA INDICATES THE STAR'S SPECTRAL TYPE, WHICH IS CORRELATED WITH ITS COLOUR AND SURFACE TEMPERATURE. THE ORDINATE SHOWS, LEFT, THE ABSOLUTE MAGNITUDE (I.E. THE INTRINSIC LUMINOSITY), AND RIGHT, THE STAR'S BRIGHTNESS COMPARED WITH THAT OF THE SUN. SIZES OF THE STARS ARE SHOWN IN TERMS OF SOLAR RADII R⊙. FINALLY, THE SUN'S PATH ACROSS THE DIAGRAM, FROM ITS ORIGIN TO ITS EVENTUAL EXTINCTION, IS SHOWN BY THE ARROWS.

| 05 | B | Spectral type | A | F | G |

| 50 000 | 30 000 | | 10 000 | 7 500 | 6 000 |

RUSSELL DIAGRAM

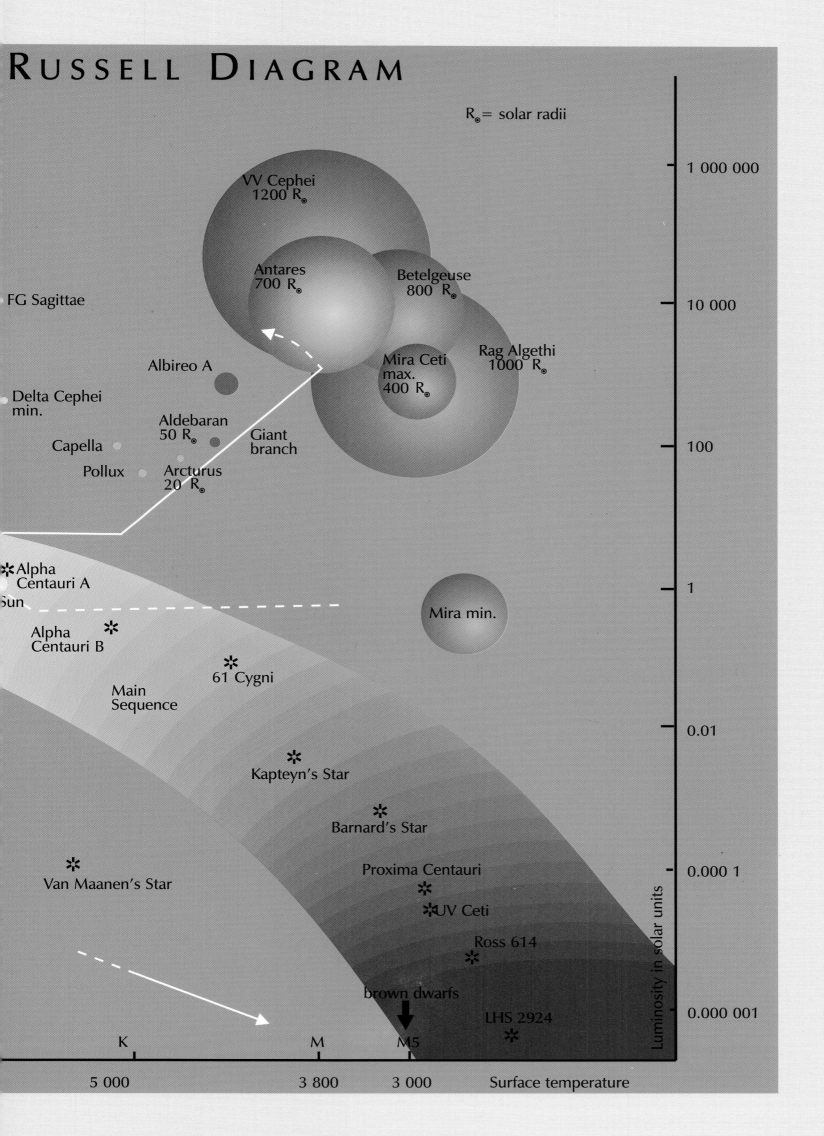

$R_{\odot}$ = solar radii

VV Cephei
1200 $R_{\odot}$

Antares
700 $R_{\odot}$

Betelgeuse
800 $R_{\odot}$

FG Sagittae

Rag Algethi
1000 $R_{\odot}$

Albireo A

Mira Ceti
max.
400 $R_{\odot}$

Delta Cephei
min.

Aldebaran
50 $R_{\odot}$

Giant
branch

Capella

Pollux

Arcturus
20 $R_{\odot}$

1 000 000

10 000

100

Alpha
Centauri A

Sun

Mira min.

1

Alpha
Centauri B

61 Cygni

Main
Sequence

0.01

Kapteyn's Star

Barnard's Star

Proxima Centauri

0.000 1

UV Ceti

Van Maanen's Star

Ross 614

brown dwarfs

LHS 2924

0.000 001

K

M

M5

Luminosity in solar units

5 000

3 800

3 000

Surface temperature

SKY

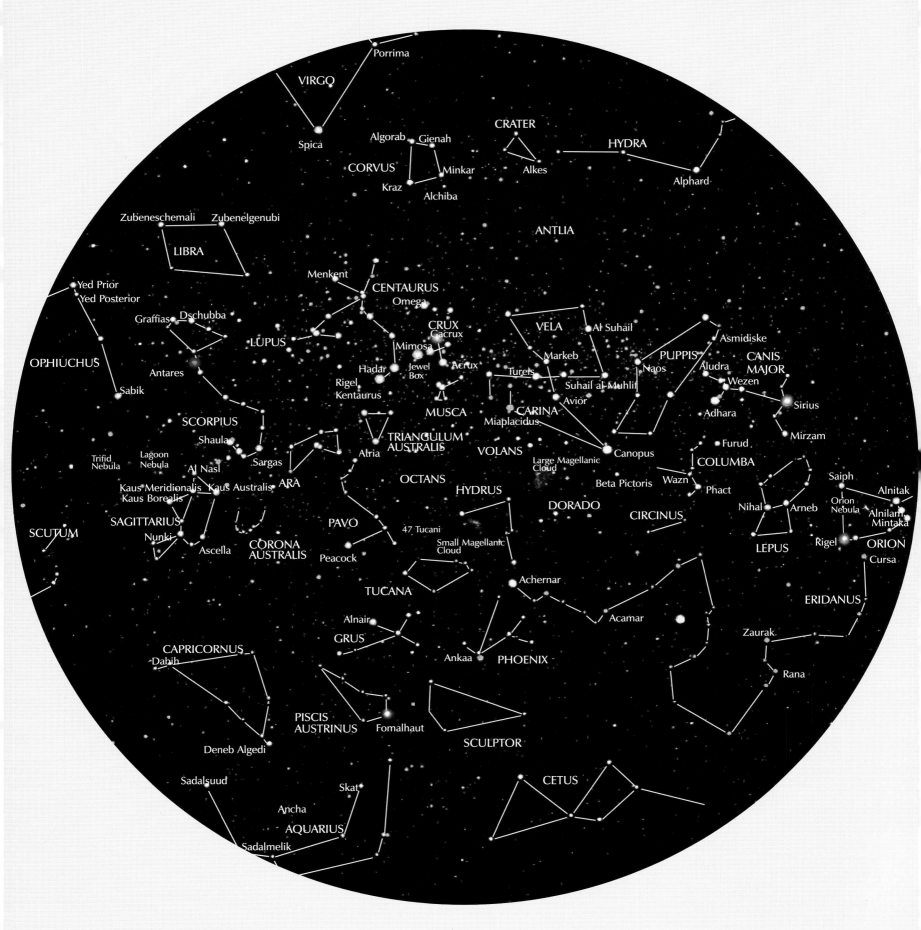

Porrima
VIRGO
Spica
CRATER
Algorab · Gienah
HYDRA
CORVUS
Minkar
Alkes
Kraz
Alchiba
Alphard

Zubeneschemali · Zubenelgenubi
ANTLIA
LIBRA

Yed Prior
Yed Posterior
Menkent
CENTAURUS
Omega
VELA
Al Suhail
Asmidiske
CANIS
MAJOR
Graffias · Dschubba
CRUX
Gacrux
Markeb
PUPPIS
Aludra
Sirius
LUPUS
Mimosa
Acrux
Turels
Naos
Wezen
Antares
Hadar
Jewel Box
Suhail al-Muhlif
Adhara
Mirzam
Sabik
Rigel Kentaurus
Avior
CARINA
Furud
SCORPIUS
MUSCA
Miaplacidus
Canopus
COLUMBA
Shaula
TRIANGULUM AUSTRALIS
VOLANS
Large Magellanic Cloud
Beta Pictoris
Wazn
Saiph
Trifid Nebula
Lagoon Nebula
Sargas
Atria
Phact
Nihal
Arneb
Alnitak
Orion Nebula
Alnitak
Al Nasl
OCTANS
HYDRUS
DORADO
CIRCINUS
Orion Nebula
Alnilam
Mintaka
Kaus Meridionalis · Kaus Australis
ARA
LEPUS
Rigel
ORION
Kaus Borealis
PAVO
47 Tucani
Cursa
SCUTUM
SAGITTARIUS
Nunki
Ascella
CORONA AUSTRALIS
Small Magellanic Cloud
Achernar
ERIDANUS
Peacock
Zaurak
TUCANA
Acamar
Alnair
CAPRICORNUS
GRUS
Ankaa
PHOENIX
Rana
Dabih
PISCIS AUSTRINUS
Fomalhaut
Deneb Algedi
SCULPTOR
Sadalsuud
Skat
CETUS
Ancha
AQUARIUS
Sadalmelik
OPHIUCHUS

SOUTHERN HEMISPHERE

CHARTS

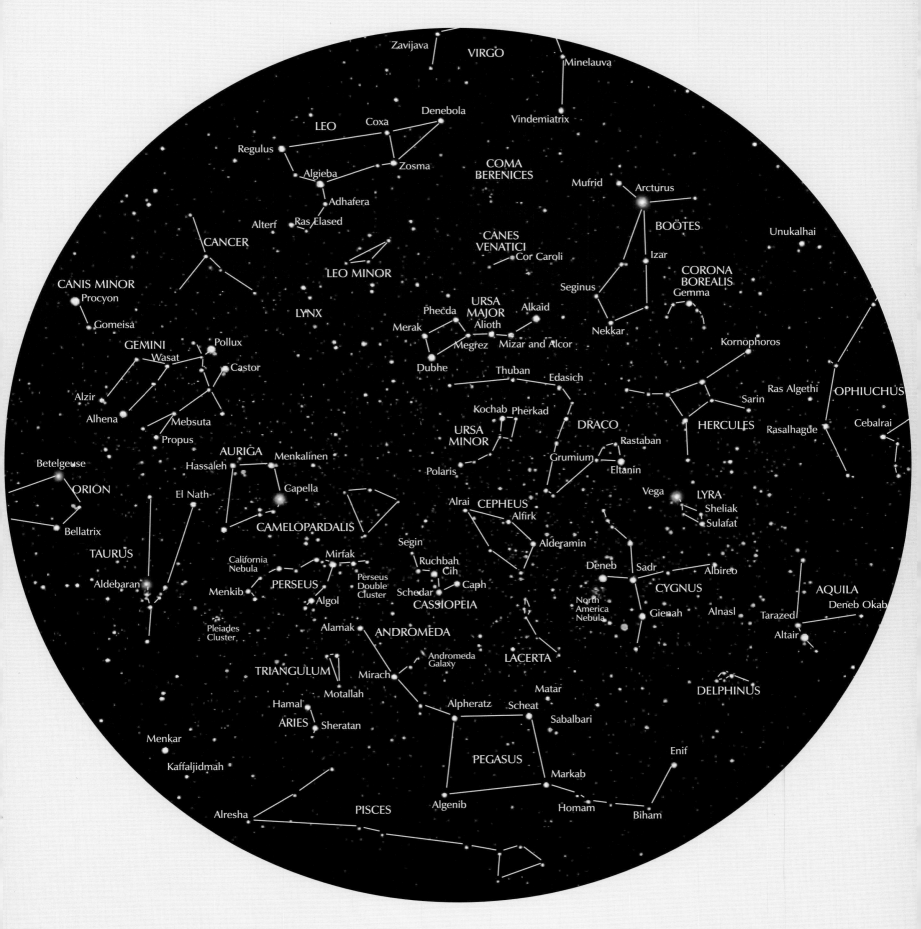

NORTHERN HEMISPHERE

THE REDSHIFT Z

A telescope is a powerful machine for looking back in time, towards the ever more distant past of the universe. Given that the speed of light is a known constant (300 000 km/s), when astronomers photograph a galaxy at a distance of 1 billion light-years (i.e. about 10 000 billion, billion kilometres), they know that the image recorded shows us the galaxy as it was 1 billion years ago. At remote distances in the universe it is, however, not possible to measure the distance of any object in light-years. On the one hand, this distance ought to be calculated using theoretical models, which take account, in particular, of the rate of expansion of the universe and its mean density; on the other hand, because of the expansion of the universe, the distance of a remote galaxy increases

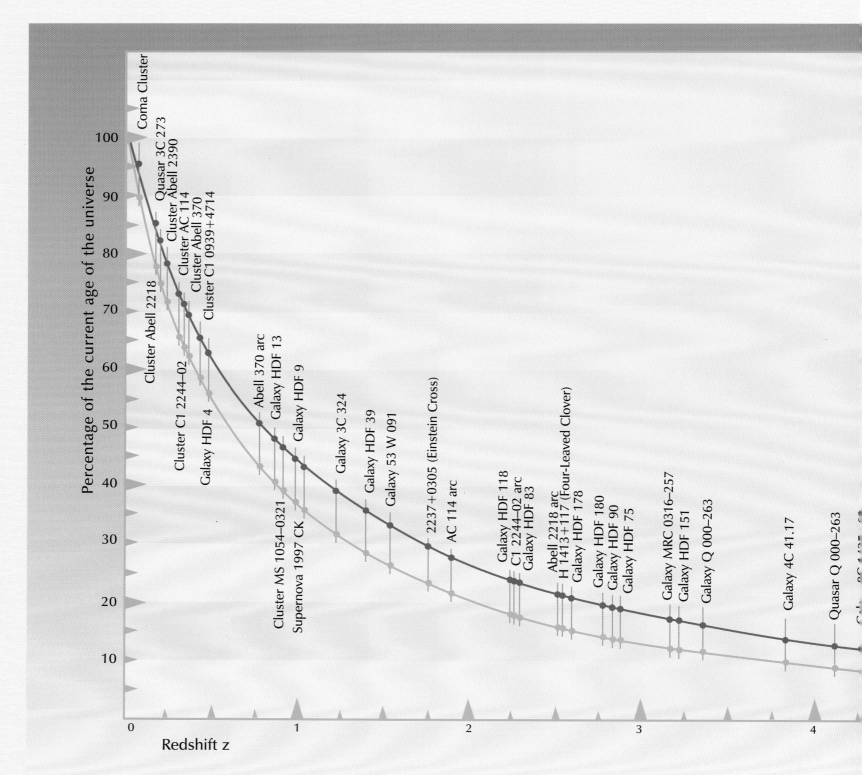

as the ray of light is travelling towards us. For this reason, in cosmology, the notion of a specific distance in space is not valid. Astronomers have substituted the idea of the look-back time, or percentage of the time back to the Big Bang. This quantity is determined from the redshift z, which is a measure of the apparent recession of galaxies in the universe, and avoids having to give a numerical value for the age of the universe. The two curves shown here indicate (on the ordinate axis) the percentage of the current age of the universe as a function of the redshift z (on the abscissa). The lower curve corresponds to the inflationary model, where the curvature parameter, $\Omega = 1$. The upper curve corresponds to a curvature parameter, $\Omega = 0.2$. To give some examples, a galaxy at $z = 1.0$ is seen as it was when the universe was 40% of its current age. If the universe is 15 billion years old, the galaxy thus appears to be 6 billion years old when we see it. At $z = 2.0$, we are looking back over 75% of the universe's past; at $z = 3.0$, 80%; at $z = 5.0$, 90%; etc. The most distant objects observed in 1998 lay at $z = 5.3$. Astronomers hope to pass $z = 6.0$ around the turn of the century.

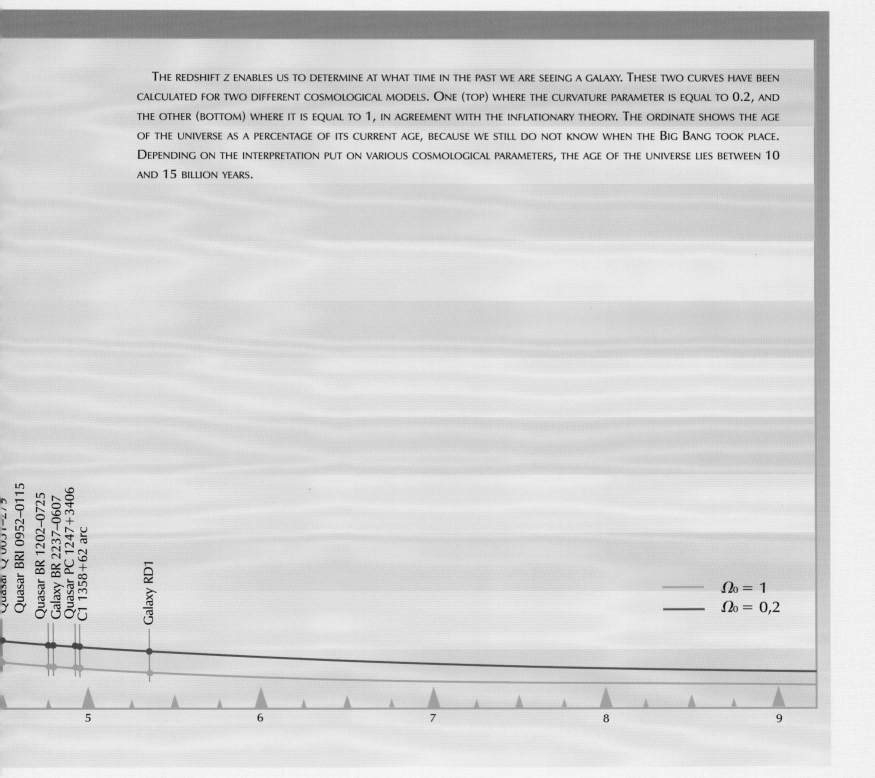

THE REDSHIFT z ENABLES US TO DETERMINE AT WHAT TIME IN THE PAST WE ARE SEEING A GALAXY. THESE TWO CURVES HAVE BEEN CALCULATED FOR TWO DIFFERENT COSMOLOGICAL MODELS. ONE (TOP) WHERE THE CURVATURE PARAMETER IS EQUAL TO 0.2, AND THE OTHER (BOTTOM) WHERE IT IS EQUAL TO 1, IN AGREEMENT WITH THE INFLATIONARY THEORY. THE ORDINATE SHOWS THE AGE OF THE UNIVERSE AS A PERCENTAGE OF ITS CURRENT AGE, BECAUSE WE STILL DO NOT KNOW WHEN THE BIG BANG TOOK PLACE. DEPENDING ON THE INTERPRETATION PUT ON VARIOUS COSMOLOGICAL PARAMETERS, THE AGE OF THE UNIVERSE LIES BETWEEN 10 AND 15 BILLION YEARS.

APPARENT FIELD AND RESOLUTION

The apparent size of astronomical objects, which all appear to be projected, at the same distance, onto the celestial sphere, are expressed in degrees (°), arcminutes (') or arcseconds ("), where 1° equals 60', and 1' equals 60". To give some idea of size, the Sun, the Moon, and the galaxy M 33 in Triangulum all appear at approximately the same size, 0.5°, or 30', or 1800". These three objects do, of course, have very different sizes, and it is the effect of perspective caused by their different distances that results in their subtending the same angle. A person with good sight is able to detect an angle of about 1' with the naked eye. It is possible to see the maria on the Moon, for example, but not the craters, the latter are too small. Telescopes are, in a manner of speaking, giant eyes: the greater the diameter of

the mirror, the better they are able to detect fine detail (described by specialists as the separation power, or the angular resolution). On Earth, where atmospheric turbulence hampers astronomers, the largest telescopes are able to resolve details that are, on average, about 1" across. On the Moon that would represent a crater 1800 m in diameter. The Hubble Space Telescope in Earth orbit, is able to obtain photographs with a resolution of 0.1": 180 m on the

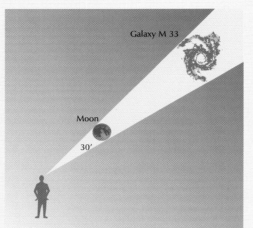

Moon, or 180 million kilometres at a distance of 30 light-years. The apparent disks of stars, on the other hand, are always less than 0.1". With the notable exception of the red supergiant Betelgeuse, whose disk has been photographed directly by the HST, stars thus never appear as anything other than bright points of light in telescopes. Under certain conditions, however, special telescopes known as interferometers are able to obtain extraordinary resolutions of approximately 0.01" to 0.0005", depending on the wavelength of observation. A figure of 0.0005" represents a detail of less than 1 m on the Moon. It is these interferometric methods that have allowed astronomers to measure the true diameters of a few tens of stars close to the Earth.

APPARENT MAGNITUDE AND ABSOLUTE MAGNITUDE

To measure and compare the apparent brightness of objects, astronomers use the apparent magnitude scale (m). This is logarithmic: each magnitude step corresponds to a change in brightness by a factor of approximately 2.5. For every 5 magnitudes the brightness changes by a factor of 100. The brightest objects have a negative apparent magnitude: −26.7 for the Sun, −12.8 for the Full Moon, −1.4 for Sirius. Vega has an apparent magnitude of 0.0, Deneb 1.25, Beta Pictoris 3.8, etc. On a clear night, a person with good eyesight can see, with the naked eye, stars of apparent magnitude 5 to 6. At the beginning of the century, the largest telescopes were able to obtain photographs of stars and galaxies of magnitude 20. In the 1950s this limit was pushed back to about magnitude 23. With electronic cameras, the best modern telescopes are capable of

detecting objects of magnitude 28. Finally, the Hubble Space Telescope, in

Apparent magnitude		Absolute Magnitude	
30 –	Limit of the HST	30 –	The Earth
28 –		28 –	
26 –		26 –	Jupiter
24 –		24 –	
22 –		22 –	
20 –		20 –	
18 –		18 –	
16 –		16 –	Proxima Centauri
14 –	Quasar 3C 273	14 –	Bernard's Star
12 –	Proxima Centauri	12 –	
10 –	Barnard's Star	10 –	
8 –	Galaxy M 87	8 –	
6 –	Andromeda Galaxy	6 –	Sun
4 –	Supernova 1987A	4 –	Altair
2 –	Deneb Altair	2 –	Sirius
0 –	Vega	0 –	Vega
-2 –	Sirius Jupiter	-2 –	
-4 –	Venus	-2 –	Betelgeuse
-6 –		-6 –	
-8 –		-8 –	Deneb
-10 –	Quarter Moon	-10 –	
-12 –	Full Moon	-12 –	
-14 –		-14 –	
-18 –		-18 –	
-18 –		-18 –	Supernova 1987A
-20 –		-20 –	Andromeda Galaxy
-22 –		-22 –	Galaxy M 87
-24 –		-24 –	
-26 –		-26 –	Quasar 3C 273
-28 –	Sun	-28 –	
-30 –		-30 –	

1995, crossed the 30th-magnitude barrier, which corresponds to objects that are 10 billionths of the brightness of the faintest object visible to the naked eye.

To be able to compare the true brightness of stars, astronomers devised the absolute magnitude scale (M). The absolute magnitude corresponds to the magnitude that the object would have if it were situated at a standard distance of 10 parsecs, i.e. about 32 light-years. The absolute magnitude of the Sun is 4.7, that of Deneb is −8.7. This means that the supergiant star in Cygnus is actually 300 000 times as luminous as the Sun. Red dwarfs have absolute magnitudes of between 15 and 20, whereas the absolute magnitude of the Andromeda Galaxy is around −21; that of M 87 is −22; those of the supergiant elliptical galaxies in the Coma Cluster, −23; and the brightest quasars are about −30.

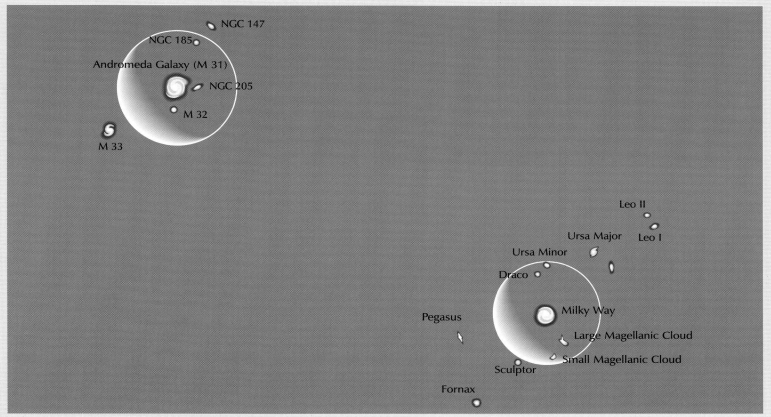

■ THE LOCAL GROUP IS DOMINATED BY TWO GIANT SPIRALS, M 31 IN ANDROMEDA AND THE MILKY WAY. THESE TWO GALAXIES ARE SURROUNDED BY A MASSIVE HALO OF INVISIBLE, DARK MATTER. IN THIS DIAGRAM, THE SIZES AND RESPECTIVE DISTANCES OF THE GALAXIES HAVE BEEN RETAINED.

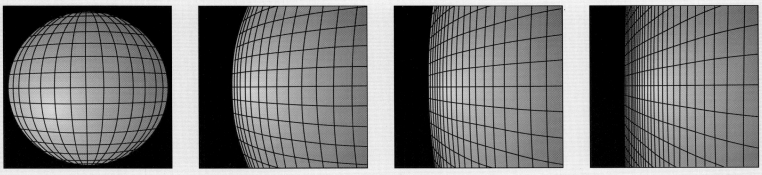

■ THE ACTION OF GRAVITATION LENSES. THE MASS OF A CLUSTER OF GALAXIES WARPS SPACE AROUND IT, AND DEFORMS AND MAGNIFIES THE IMAGES OF GALAXIES THAT LIE IN THE BACKGROUND. DEPENDING ON THE PRECISE ALIGNMENT OF THE GALAXY, CLUSTER AND EARTH, THE IMAGES MAY APPEAR AS A CROSS, CIRCLES, OR ARCS.

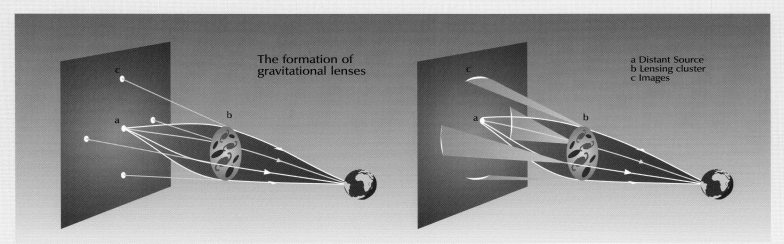

■ THE THEORY OF INFLATION. IN AN INFINITESIMAL FRACTION OF A SECOND AFTER THE BIG BANG, THE SIZE OF THE UNIVERSE INCREASED BY TENS OF ORDERS OF MAGNITUDE. THIS IS WHY THE CURVATURE OF THE UNIVERSE IS NOW UNDETECTABLE: THE CURVATURE PARAMETER, IS EQUAL TO 1.

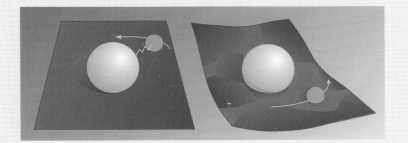

■ TWO REPRESENTATIONS OF GRAVITATION. IN NEWTON'S MODEL (LEFT) IT IS A FORCE; IN THAT OF EINSTEIN'S (RIGHT) IT IS AN EFFECT OF THE CURVATURE OF SPACE.

THE BRIGHTEST STARS IN THE SKY, AS

STAR	BAYER DESIG.		m	M	L (solar)	P (")	D (l-y)
ANDROMEDA							
Alpheratz	α	And	2.07	−0.30	110	34	96
Mirach	β	And	2.07	−1.86	500	16	200
Alamak	γ¹	And	2.10	−3.08	1000	9	400
	δ	And	3.27	0.81	40	32	100
AQUARIUS							
Sadalsuud	β	Aqr	2.9	−3.47	2000	5	700
Sadalmelik	α	Aqr	2.95	−3.88	3000	4	800
Skat	δ	Aqr	3.27	−0.18	100	20	160
AQUILA							
Altair	α	Aql	0.76	2.20	11.2	194	16.8
Tarazed	γ	Aql	2.72	−3.03	1000	7	500
	ζ	Aql	2.99	0.96	35	39	84
	θ	Aql	3.24	−1.48	300	11	300
Deneb el Okab	δ	Aql	3.36	2.43	9	65	50
	λ	Aql	3.43	0.51	50	26	130
ARA							
	β	Ara	2.84	−3.49	2000	5	700
	α	Ara	2.84	−1.51	300	13	250
	ζ	Ara	3.12	−3.11	1000	6	500
	γ	Ara	3.31	−4.40	5000	3	1000
ARIES							
Hamal	α	Ari	2.01	0.48	54	49	67
Sheratan	β	Ari	2.64	1.33	25	55	59
AURIGA							
Capella	α	Aur	0.08	−0.48	130	77	42
Menkalinan	β	Aur	1.9	−0.10	90	40	82
	θ	Aur	2.65	−0.98	210	19	170
Hassaleh	ι	Aur	2.69	−3.29	2000	6	500
Almaz	ε	Aur	3.03	−5.95	20 000	2	2000
Hoedus	η	Aur	3.18	−0.96	210	15	220
BOÖTES							
Arcturus	α	Boo	−0.05	−0.31	110	89	37
Izar	ε	Boo	2.35	−1.69	400	16	200
Muphrid	η	Boo	2.68	2.41	9.2	88	37
Haris, Seginus	γ	Boo	3.04	0.96	35	38	86
	δ	Boo	3.46	0.69	40	28	120
Nekhar	β	Boo	3.49	−0.64	150	15	220
CANCER							
Altarf	β	Cnc	3.53	−1.22	300	11	300
CAPRICORNUS							
Dened Algedi	δ	Cap	2.85	2.49	8.6	85	38
Dabih	β	Cap	3.05	−2.07	600	9	400
CANES VENATICI							
Cor Caroli	α²	CVn	2.89	0.25	70	30	110
CANIS MAJOR							
Sirius	α	CMa	−1.44	1.45	22	379	8.61
Adhara	ε	CMa	1.5	−4.10	4000	8	400
Wezen	δ	CMa	1.83	−6.87	50 000	2	2000
Mirzam	β	CMa	1.98	−3.95	3000	7	500
Aludra	η	CMa	2.45	−7.51	100 000	1	3000
Furud	ζ	CMa	3.02	−2.05	600	10	300
	o²	CMa	3.02	−6.46	30 000	1	3000
	σ	CMa	3.49	−4.37	5000	3	1000
	κ	CMa	3.5	−3.42	2000	4	800
CANIS MINOR							
Procyon	α	CMi	0.4	2.68	7.2	286	11.4
Gomeisa	β	CMi	2.89	−0.70	160	19	170
CARINA							
Canopus	α	Car	−0.62	−5.53	14 000	10	300
Miaplacidus	β	Car	1.67	−0.99	210	29	110
Avior	ε	Car	1.86	−4.58	6000	5	700
Tureis	ι	Car	2.21	−4.42	5000	5	700
	θ	Car	2.74	−2.91	1000	7	500
	μ	Car	2.92	−5.56	10 000	2	2000
	ω	Car	3.29	−1.99	500	9	400
HR 4140	ρ	Car	3.3	−2.62	900	7	500
HR 4050	θ	Car	3.39	−3.38	2000	4	800
HR 3659	α	Car	3.43	−2.11	600	8	400
	χ	Car	3.46	−1.91	500	8	400
CASSIOPEIA							
Cih	g	Cas	2.15	−4.22	4000	5	700
Schedar	α	Cas	2.24	−1.99	500	14	230
Caph	β	Cas	2.28	1.17	29	60	54
Ruchbah	δ	Cas	2.66	0.24	70	33	99
Segin	ε	Cas	3.35	−2.31	700	7	500
Achird	η	Cas	3.46	4.59	1.24	168	19.4
CENTAURUS							
Rigil Kent, Toliman	α¹	Cen	−0.01	4.34	1.56	742	4.40
Agena, Hadar	β	Cen	0.61	−5.42	10 000	6	500
	α²	Cen	1.35	5.70	0.44	742	4.40
Menkent	θ	Cen	2.06	0.70	44	54	60
	γ	Cen	2.2	−0.81	180	25	130
	ε	Cen	2.29	−3.02	1000	9	400
	η	Cen	2.33	−2.55	900	11	300
	z	Cen	2.55	−2.81	1100	8	400
	δ	Cen	2.58	−2.84	1200	8	400
	ι	Cen	2.75	1.48	22	56	58
	λ	Cen	3.11	−2.39	800	8	400
	κ	Cen	3.13	−2.96	1000	6	500
	υ	Cen	3.41	−2.41	800	7	500
	μ	Cen	3.47	−2.57	900	6	500
CEPHEUS							
Alderamin	α	Cep	2.45	1.58	20	67	49
Alrai	γ	Cep	3.21	2.51	8.4	72	45
Alfirk	β	Cep	3.23	−3.08	1000	5	700
	ζ	Cep	3.39	−3.35	2000	4	800
	η	Cep	3.41	2.63	7.5	70	47
	ι	Cep	3.5	0.76	40	28	120
CETUS							
Diphda	β	Cet	2.04	−0.30	110	34	96
Menkar	α	Cet	2.54	−1.61	400	15	220
	η	Cet	3.46	0.67	50	28	120
Kaffaljidhmah	γ	Cet	3.47	1.47	22	40	82
	t	Cet	3.49	5.68	0.45	274	11.9
CIRCINUS							
	α	Cir	3.18	2.11	12	61	53
COLUMBA							
Phact	α	Col	2.65	−1.93	500	12	270
Wazn	β	Col	3.12	1.02	33	38	86
CORONA BOREALIS							
Gemma	α	CrB	2.22	0.42	58	44	74
CORVUS							
Gienah	γ	Crv	2.58	−0.94	200	20	160
Kraz	β	Crv	2.65	−0.51	140	23	140
Algorab	δ	Crv	2.94	0.79	41	37	88
Minkar	ε	Crv	3.02	−1.82	500	11	300
CRUX							
Acrux	α¹	Cru	0.77	−4.19	4000	10	300
Mimosa	β	Cru	1.25	−3.92	3000	9	400
Gacrux	γ	Cru	1.59	−0.56	140	37	88
	δ	Cru	2.79	−2.45	800	9	400
CYGNUS							
Deneb	α	Cyg	1.25	−8.73	300 000	1	3000
Sadr	γ	Cyg	2.23	−6.12	20 000	2	2000
Gienah	ε	Cyg	2.48	0.76	42	45	72
	δ	Cyg	2.86	−0.74	170	19	170
Albireo	β¹	Cyg	3.05	−2.31	700	8	400
	ζ	Cyg	3.21	−0.12	90	22	150
DORADO							
	α	Dor	3.3	−0.36	120	19	170
DRACO							
Eltanin	γ	Dra	2.24	−1.04	220	22	150
	η	Dra	2.73	0.58	50	37	88
Rastaban	β	Dra	2.79	−2.43	800	9	400
Nodus II	δ	Dra	3.07	0.63	50	33	99
Nodus I	ζ	Dra	3.17	−1.92	500	10	300
Edasich	ι	Dra	3.29	0.81	40	32	100
ERIDANUS							
Achernar	α	Eri	0.45	−2.77	1100	23	140
Cursa	β	Eri	2.78	0.60	49	37	88
Acamar	θ¹	Eri	2.88	−0.59	150	20	160
Zaurak	γ	Eri	2.97	−1.19	300	15	220
Rana	δ	Eri	3.52	3.74	2.7	111	29.4
GEMINI							
Pollux	β	Gem	1.16	1.09	31	97	34
Castor	α	Gem	1.58	0.59	49	63	52
Alhena	γ	Gem	1.93	−0.60	150	31	110
Tejat posterior	μ	Gem	2.87	−1.39	300	14	230
Mebsuta	e	Gem	3.06	−4.15	4000	4	800
Propus	η	Gem	3.31	−1.84	500	9	400
Alzir	x	Gem	3.35	2.13	12	57	57
Wasat	δ	Gem	3.5	2.22	11	55	59
GRUS							
Alnair	α	Gru	1.73	−0.73	170	32	100
	β	Gru	2.07	−1.52	300	19	170
	γ	Gru	3	−0.97	210	16	200
	e	Gru	3.49	0.49	50	25	130
HERCULES							
Kornephoros	β	Her	2.78	−0.50	130	22	150
Ras Algethi	α¹	Her	2.78	−2.57	900	9	400
	ζ	Her	2.81	2.64	7.4	93	35
Sarin	δ	Her	3.12	1.21	28	42	78
	π	Her	3.16	−2.10	600	9	400
HYDRA							
Alphard	α	Hya	1.99	−1.69	400	18	180
	γ	Hya	2.99	−0.05	90	25	130
	ζ	Hya	3.11	−0.21	100	22	150
	υ	Hya	3.11	−0.03	90	24	140
	π	Hya	3.25	0.79	41	32	100
	ε	Hya	3.38	0.29	60	24	140
	ζ	Hya	3.54	0.55	50	25	130
HYDRUS							
	β	Hyi	2.82	3.45	3.5	134	24.3
	α	Hyi	2.86	1.16	29	46	71
	γ	Hyi	3.26	−0.83	180	15	220
INDUS							
	α	Ind	3.11	0.65	50	32	100
LEO							
Regulus	α	Leo	1.36	−0.52	140	42	78
Algieba	γ¹	Leo	2.01	−0.92	200	26	130
Denebola	β	Leo	2.14	1.92	14	90	36
Zozma	δ	Leo	2.56	1.32	25	57	57
Ras Elased	ε	Leo	2.97	−1.46	300	13	250
Coxa	θ	Leo	3.33	−0.35	120	18	180
Adhafera	ζ	Leo	3.43	−1.08	200	13	250
	η	Leo	3.48	−5.60	10 000	2	2000
Subra	o	Leo	3.52	0.43	60	24	140

MEASURED BY THE HIPPARCOS SATELITE

Name							
LEPUS							
Arneb	α	Lep	2.58	−5.40	10 000	3	1000
Nihal	β	Lep	2.81	−0.63	150	20	160
	ε	Lep	3.19	−1.02	200	14	230
	μ	Lep	3.29	−0.47	130	18	180
LIBRA							
Zubeneschemali	β	Lib	2.61	−0.84	180	20	160
Zubenelgenubi	α2	Lib	2.75	0.88	38	42	78
	σ	Lib	3.25	−1.51	300	11	300
LUPUS							
	α	Lup	2.3	−3.83	3000	6	500
	β	Lup	2.68	−3.35	2000	6	500
	γ	Lup	2.8	−3.40	2000	6	500
	δ	Lup	3.22	−2.75	1000	6	500
	ε	Lup	3.37	−2.58	900	6	500
	ζ	Lup	3.41	0.65	50	28	110
	η	Lup	3.42	−2.48	800	7	500
LYNX							
	α	Lyn	3.14	−1.02	220	15	220
LYRA							
Vega	α	Lyr	0.03	0.58	50	129	25.3
Sulafat	γ	Lyr	3.25	−3.20	2000	5	700
Sheliak	β	Lyr	3.52	−3.64	2000	4	800
MUSCA							
	α	Mus	2.69	−2.17	600	11	300
	β	Mus	3.04	−1.86	500	10	300
OPHIUCHUS							
Rasalhague	α	Oph	2.08	1.30	26	70	47
Sabik	η	Oph	2.43	0.37	60	39	84
	ζ	Oph	2.54	−3.20	2000	7	500
Yed prior	δ	Oph	2.73	−0.86	190	19	170
Cebalrai	β	Oph	2.76	0.76	42	40	82
	κ	Oph	3.19	1.09	31	38	86
Yed posterior	ε	Oph	3.23	0.64	50	30	110
	ε	Oph	3.27	−2.92	1000	6	500
	υ	Oph	3.32	−0.03	90	21	160
ORION							
Rigel	β	Ori	0.18	−6.69	40 000	4	800
Betelgeuse	α	Ori	0.45	−5.14	10 000	8	400
Bellatrix	γ	Ori	1.64	−2.72	1000	13	250
Alnilam	ε	Ori	1.69	−6.38	30 000	2	2000
Alnitak	ζ	Ori	1.74	−5.26	10 000	4	800
Saiph	κ	Ori	2.07	−4.65	6000	5	700
Mintaka	δ	Ori	2.25	−4.99	10 000	4	800
	ι	Ori	2.75	−5.30	10 000	2	2000
	π3	Ori	3.19	3.67	2.9	125	26.1
	η	Ori	3.35	−3.86	3000	4	800
Heka	λ	Ori	3.39	−4.16	4000	3	1000
PAVO							
Peacock	α	Pav	1.94	−1.81	400	18	180
	β	Pav	3.42	0.29	60	24	140
PEGASUS							
Enif	ε	Peg	2.38	−4.19	4000	5	700
Scheat	β	Peg	2.44	−1.49	300	16	200
Markab	α	Peg	2.49	−0.67	160	23	140
Algenib	γ	Peg	2.83	−2.22	700	10	300
Matar	η	Peg	2.93	−1.16	200	15	220
Homam	ζ	Peg	3.41	−0.62	150	16	200
Sadalbari	μ	Peg	3.51	0.74	40	28	120
Biham	θ	Peg	3.52	1.16	29	34	96
PERSEUS							
Mirfak	α	Per	1.79	−4.50	5000	6	500
Algol	β	Per	2.09	−0.18	100	35	93
Menkib	ζ	Per	2.84	−4.55	10 000	3	1000
	ε	Per	2.9	−3.19	2000	6	500
	γ	Per	2.91	−1.57	400	13	25
	δ	Per	3.01	−3.04	1000	6	500
	ρ	Per	3.32	−1.67	400	10	300
PHEONIX							
Ankaa	α	Phe	2.4	0.52	52	42	78
	β	Phe	3.32	−0.60	150	16	200
	γ	Phe	3.41	−0.87	190	14	230
PICTOR							
	α	Pic	3.24	0.83	39	33	99
Beta Pictoris	β	Pic	3.85	2.42	7	51	63
PISCIS AUSTRINUS							
Fomalhaut	α	PsA	1.17	1.74	17	130	25.1
PUPPIS							
Naos	ζ	Pup	2.21	−5.95	20 000	2	2000
	π	Pup	2.71	−4.92	10 000	3	1000
	ρ	Pup	2.83	1.41	23	52	63
	τ	Pup	2.94	−0.80	180	18	180
	υ	Pup	3.17	−2.39	800	8	400
	σ	Pup	3.25	−0.51	140	18	180
Asmidiske	ξ	Pup	3.34	−4.74	10 000	2	2000
RETICULUM							
	α	Ret	3.33	−0.17	100	20	160
SAGITTA							
	γ	Sge	3.51	−1.11	200	12	270
SAGITTARIUS							
Kaus Australis	ε	Sgr	1.79	−1.44	320	23	140
Nunki	σ	Sgr	2.05	−2.14	600	15	220
Ascella	ζ	Sgr	2.6	0.42	60	37	88
Kaus Meridionalis	δ	Sgr	2.72	−2.14	600	11	300
Kaus Borealis	λ	Sgr	2.82	0.95	35	42	78

Name							
Al Nasl	p	Sgr	2.88	−2.77	1000	7	500
	γ	Sgr	2.98	0.63	50	34	96
	η	Sgr	3.1	−0.20	100	22	150
	j	Sgr	3.17	−1.08	200	14	230
	t	Sgr	3.32	0.48	50	27	120
Sulafat	ξ2	Sgr	3.52	−1.77	400	9	400
SCORPIUS							
Antares	α	Sco	1.06	−5.28	10 000	5	700
Shaula	λ	Sco	1.62	−5.05	10 000	5	700
Sargas	θ	Sco	1.86	−2.75	1100	12	270
Dschubba	δ	Sco	2.29	−3.16	2000	8	400
	ε	Sco	2.29	0.78	41	50	65
	κ	Sco	2.39	−3.38	2000	7	500
Akrab, Graffias	β1	Sco	2.56	−3.50	2000	6	500
	υ	Sco	2.7	−3.31	2000	6	500
	τ	Sco	2.82	−2.78	1100	8	400
	π	Sco	2.89	−2.85	1000	7	500
Al Niyat	σ	Sco	2.9	−3.86	3000	4	800
	ι1	Sco	2.99	−5.71	20 000	2	2000
	μ1	Sco	3	−4.01	3000	4	800
	η	Sco	3.32	1.61	19	46	71
HR 6630	γ	Sco	3.19	0.24	70	26	130
SERPENS							
Unukalhai	α	Ser	2.63	0.87	38	45	72
	η	Ser	3.23	1.84	16	53	62
	μ	Ser	3.54	0.14	70	21	160
TAURUS							
Aldebaran	α	Tau	0.87	−0.63	150	50	65
Elnath	β	Tau	1.65	−1.37	300	25	130
Alcyone	η	Tau	2.85	−2.41	800	9	400
	ζ	Tau	2.97	−2.56	900	8	400
	θ2	Tau	3.4	0.10	80	22	150
	λ	Tau	3.41	−1.87	500	9	400
Ain	ε	Tau	3.53	0.15	70	21	160
TELESCOPIUM							
	α	Tel	3.49	−0.93	200	13	250
TRIANGULUM							
	β	Tri	3	0.09	80	26	130
Motallah	α	Tri	3.42	1.95	14	51	64
TRIANGULUM AUSTRALE							
Atria	α	TrA	1.91	−3.62	2000	8	400
	β	TrA	2.83	2.38	9	81	40
	γ	TrA	2.87	−0.87	190	18	180
TUCANA							
	α	Tuc	2.87	−1.05	220	16	200
URSA MAJOR							
Alioth	ε	UMa	1.76	−0.21	100	40	82
Dubhe	α	UMa	1.81	−1.08	230	26	130
Alkaid	η	UMa	1.85	−0.60	150	32	100
Mizar	ζ	UMa	2.23	0.33	60	42	78
Merak	β	UMa	2.34	0.41	60	41	80
Phecda	γ	UMa	2.41	0.36	60	39	84
	γ	UMa	3	−0.27	110	22	150
Tania Australis	μ	UMa	3.06	−1.35	300	13	250
Talitha	ι	UMa	3.12	2.29	10	68	48
	θ	UMa	3.17	2.52	8.3	74	44
Megrez	δ	UMa	3.32	1.33	25	40	82
Muscida	o	UMa	3.35	−0.40	120	18	180
Tania Borealis	λ	UMa	3.45	0.38	60	24	140
Alula Borealis	υ	UMa	3.49	−2.07	600	8	400
URSA MINOR							
Polaris	α	UMi	1.97	−3.64	2000	8	400
Kochab	β	UMi	2.07	−0.87	190	26	130
Pherkad	γ	UMi	3	−2.84	1000	7	500
VELA							
Suhail al Muhlif	γ	Vel	1.75	−5.31	10 000	4	800
	δ	Vel	1.93	−0.01	90	41	80
Al Suhail	λ	Vel	2.23	−3.99	3000	6	500
Markab	κ	Vel	2.47	−3.62	2000	6	500
	μ	Vel	2.69	−0.06	90	28	120
HR 3806	υ	Vel	3.16	−1.15	200	14	230
	φ	Vel	3.52	−5.34	10 000	2	2000
VIRGO							
Spica	α	Vir	0.98	−3.55	2000	12	270
Porrima	γ	Vir	2.74	2.38	9.5	85	38
Vindemiatrix	ε	Vir	2.85	0.37	60	32	100
Heze	ζ	Vir	3.38	1.62	19	45	72
Minelauva	δ	Vir	3.39	−0.57	140	16	200

THE HIPPARCOS MISSION

The Hipparcos mission is one of the most ambitious ever undertaken in the history of astronomy. It was officially proposed by the European Space Agency in 1980. The aim was to measure precisely the positions, proper motions, and apparent magnitudes of all the stars in the solar neighbourhood. The Hipparcos satellite (an acronym for High Precision Parallax Collecting Satellite) operated between 11 September 1989 and 15 August 1993. After four years of data reduction, ESA published the 17 volumes of this catalogue of fundamental stars, the most accurate ever compiled, in May 1997. A total of 118 218 stars are catalogued to a precision of 1 arcsecond, and an additional 1 050 000 stars to an accuracy of between 0.003 and 0.5 arcsecond.

Thanks to these data, astronomers have, for the first time, a giant three-dimensional map of our galactic environment, within a sphere 6000 light-years in diameter. This map is also to some degree a map in both space and time, because the satellite followed the individual motion of each star over the course of the mission, which also allows astronomers to obtain an overall picture of the motion of stars in the Galaxy over the course of time. The Hipparcos catalogue will thus serve as a reference for decades to come. Practically all branches of astronomy will be affected, some more closely than others, by the results of this mission, particularly stellar astrophysics and cosmology. This is because the Hipparcos catalogue is not a simple static atlas, showing the state of our galactic neighbourhood. Its measurements are real keys to the understanding of some of today's most disputed astrophysical phenomena. In particular, our knowledge, before

Hipparcos, of the rarest stars, the blue and red supergiants, the giant pulsating stars – which are the most interesting in trying to understand stellar evolution – was very patchy. In the Galaxy these extremely rare stars are sparsely distributed, and none is found in the immediate vicinity of the Sun. As a result, it was impossible for astronomers, using ground-based telescopes, which are handicapped by atmospheric turbulence, to determine their distances precisely. From an accurate measurement of distance and an accurate measurement of the apparent magnitude of an object, however, it is immediately possible to determine its true brightness and, with the aid of the theories of stellar nucleosynthesis, its size, its mass, its energy output and its age.

The Hipparcos mission was based on the principle of measuring trigonometrical parallax. When one observes a nearby object, say a church tower, in front of a distant background such as a range of hills, and if one moves slightly, the nearby object seems to move relative to the background. This effect, known as 'parallax' also exists for stars, but it is extremely small. At first glance all the stars appear at the same distance, projected onto the celestial sphere. Some of them are, however, very close to us, say at distances of a few light-years, whereas others are some 1000 or 10 000 times as far away. When these stars are observed from Earth at six-monthly intervals – i.e. at two extremes of the Earth's orbit – the apparent movement (known as the parallax), of nearby stars relative to more distant ones becomes apparent. Nearby stars describe tiny loops in front of the distant background stars, these loops being, in effect, the Earth's orbit projected against the sky.

Knowing the size of the base (300 million kilometres) and the apparent displacement of a star, it is easy to calculate its true distance. These angles, expressed in arcseconds (") are minute. All the parallaxes measured by Hipparcos lie between 0.742" and 0.001". One thousandth of an arcsecond represents the size of a golf ball on top of the Empire State Building in New York, as seen from Europe. Or something the size of a man, seen from the Moon. The average accuracy of the Hipparcos measurements is 0.001". That implies that the closer the star, the larger its parallax, and the higher the degree of accuracy. On the other hand, the more the distance increases, the smaller the parallax becomes, and the lower the accuracy. Beyond 0.001", which corresponds to a distance of 1000 parsecs – a parsec is the distance at which the radius of the Earth's orbit subtends an angle of 1"; 1 parsec corresponds to 3.26 light-years – Hipparcos is no longer able to detect the slight angular changes in the position of a star.

The satellite's mission consisted of simultaneously observing two regions of the sky, 58° apart, in order to measure the angular distances between the stars, taken in pairs, extremely accurately. Hipparcos therefore behaved like a pair of compasses, tracing great circles on the sky, sweeping past each star a hundred times, each time with a different orientation. After the mission, the analysis consisted of reconstructing the whole map from the hundreds of millions of pairs of measurements. The table on the preceding two pages gives, for the first time, the exact details of the 300 brightest stars in the sky.

GLOSSARY

3-K background radiation (cosmic background radiation, background radiation) Electromagnetic radiation coming from all directions in the sky, and which has the properties of thermal radiation from a black body at a temperature that is close to 3 K. Discovered in 1965 by A. Penzias and R. Wilson, it is easily explained by the theory of the Big Bang, which it decisively confirmed.

accretion Capture of material by a body by gravitational attraction. Accretion disk: a disk-shaped region surrounding a white dwarf, a neutron star, or a black hole, through which the material spirals before finally falling onto the compact object.

Big Bang The event that, according to the cosmological models currently generally accepted marked the origin of space and time and thus of the universe as it is observed today. Initially the temperature and density of the universe would have been extremely high, but these would have decreased as the expansion took place. Three observational facts may be advanced as evidence in favour of the Big-Bang theory. The first is the general recession of the galaxies, which is an expression of the expansion of space. The second is the sky's background radiation, which is most easily interpreted as the fossil radiation remaining from the initial extremely hot state. The experimental evidence is the abundance, in the universe, of the lightest chemical elements, in particular, hydrogen and helium. The values observed imply that the universe passed through a primordial phase that was exceptionally hot and dense, during which the nuclear reactions occurred that produced the nuclei of the light atoms (known as the period of primordial nucleosynthesis).

binary star A pair of stars linked by their mutual gravitational attraction.

black hole A region of the universe that undergoes irreversible gravitational collapse, and whose gravitational field is so intense that nothing, not even light, can escape.

brown dwarf An astronomical object whose mass is estimated to lie between 17 and 80 times that of Jupiter, but which has remained too cool for it to be the site of nuclear reactions. Brown dwarfs may be one component of the missing mass in the universe.

Cepheid A type of pulsating star of which δ Cephei is prototype. The brightness of each star varies with a well-determined period. Overall, the range may lie between one day and several weeks.

closed universe A model of the universe that contains a sufficient quantity of matter to halt the expansion and cause its contraction at a specific epoch in the future.

cluster of galaxies A group of galaxies that are bound together by their mutual gravitation.

curvature of the universe A geometrical property of space-time, according to the theory of relativity, and the most obvious expression of which is gravitation. Physical phenomena occur in four-dimensional space-time. The presence of matter within this space-time creates a curvature which is greater, the greater the density of matter that is present.

dark matter (see **missing mass**)

deceleration parameter A number that measures the rate at which the expansion of the universe is decreasing as a result of the mutual gravitational attraction of the mass contained within it.

Einstein effect A term that is used indiscriminately for two phenomena caused by gravitation and predicted by general relativity: the deviation of rays of light; or the redshift produced in radiation that passes close to a massive body before reaching the Earth.

Einstein ring A circular image of a distant celestial point source, which is created when a concentration of mass lying on the line of sight acts as a gravitational lens. (Predicted by the theory of general relativity, this shape is observed for certain quasars whose radiation is deviated by closer, intervening galaxies.)

expansion of the universe A phenomenon predicted on the basis of the theory of general relativity by W. De Sitter in 1917, and then by A. Friedmann in 1922 and G. Lemaître in 1927, and according to which the various galaxies in the universe are receding from one another at a velocity proportional to their respective mutual distances.

galactic halo A spherical volume of space around a galaxy, populated by old stars occurring in globular clusters, which themselves orbit the centre of the galaxy.

galaxy A vast collection of stars and interstellar matter, isolated in space, but which is maintained as a single coherent object by its gravity.

giant A type of star that is about 100 times as luminous as the Sun, and of greater radius, but less dense. When hydrogen is exhausted in the core of a star, the core contracts, causing hydrogen to burn in layers farther from the centre, while the stellar envelope expands. In this phase the star is described as being a *red giant*.

globular cluster A spherical concentration of thousands or tens of thousands of stars, bound together by their mutual gravitation, all of similar age, and belonging to the oldest component of a galaxy.

gravitation One of the four fundamental interactions in physics, which manifests itself as attractive forces between all bodies possessing mass. Newton first formulated the law expressing this: for two point masses m and m', separated by a distance r, attract one another with a force, directed along the line between them, and whose magnitude is proportional to the two masses and inversely proportional to the square of the distance.

The theory of relativity integrates gravitation in a geometrical framework, in which a gravitational field is interpreted as a curvature of space-time caused by the presence of mass. Gravitation is the predominant force governing the evolution of the material within the universe.

gravitational lens A massive celestial object, such as a galaxy, that causes a change in the appearance of more distant objects situated along the same line of sight. The theory of general relativity predicts that light should be affected when it encounters an intense gravitational field, and that it may then be deflected or amplified. Concentrations of mass in the universe may thus act as lenses for rays of light passing through them. This phenomenon is illustrated by certain quasars, discovered since 1979, several images of which are observed, because the path of their light is curved by the gravitational field of closer galaxies that lie in the same direction.

group (of galaxies) A small cluster of galaxies that typically contains one or two dozen members.

Hertzsprung-Russell Diagram (abbr. HRD) A diagram devised at the beginning of the 20th century by E. Hertzsprung and perfected by H.N. Russell, who established a classification of stars according to their spectral type and luminosity.

Hubble Constant A factor expressing the relationship between the velocity of recession of galaxies and their distance, according to the Hubble Law. Designated H_0, this parameter expresses the rate of the universe's expansion with time, and plays a significant role in cosmological models.

Hubble Law An empirical law expressed by Hubble in 1929, in which galaxies have an apparent velocity of recession that is proportional to their distance. It is interpreted as a consequence of the expansion of the universe.

inflation According to certain cosmologists, an extremely rapid phase of expansion at an exponential, rather than linear rate, that occurred in the universe a fraction of a second after the Big Bang.

interstellar medium The extremely diffuse material (gas and dust) that lies between the stars in a galaxy.

isotropic Said of properties that are the same in all directions (in speaking of radiation, a medium, etc.)

light-year A unit of distance (symbol ly) equal to the distance covered by light in one year in a vacuum, and approximately 9.461×10^{12} km.

Local Group A small cluster of galaxies to which our Galaxy belongs. It comprises some thirty members, including the Magellanic Clouds, and the galaxy M 31 in Andromeda. It lies at the edge of a far greater cluster of galaxies, known as the Local Supercluster.

Local Supercluster The supercluster of which the Galaxy is part.

magnitude A number that indicates the apparent brightness (apparent magnitude) or intrinsic luminosity (absolute magnitude) of an object. On the magnitude scale, the smallest numbers correspond to the brightest objects.

main sequence The strip that extends diagonally across the Hertzsprung-Russell diagram from upper left to bottom right, and along which 90% of all stars lie, and which, like the Sun, derive their energy from the fusion of hydrogen into helium.

megaparsec (Mpc) The unit of distance used in extragalactic astronomy, equal to one million parsecs.

missing mass (or dark matter) Dark material that is thought to exist in the universe. The assumption that the missing mass exists is based on the dynamical behaviour of clusters of galaxies, and on dynamical studies of isolated galaxies. Analysis of both galaxies and clusters of galaxies indicates that these systems contain much more matter than their visible contents suggest. The solution to this problem of the 'hidden' mass affects the question of the eventual fate of the universe. The presence of missing mass is also invoked in theories that attempt to explain the formation of galaxies following the Big Bang. It might consist of low-mass stars (less than a few tenths of a solar mass), in particular, of brown dwarfs, or of large clouds of cold, molecular gas. According to another theory, it might consist of massive subatomic particles that interact only weakly with ordinary matter (and are thus difficult to detect).

nebula A cloud of interstellar gas and dust. The formation of new stars takes place within immense molecular clouds – clouds of interstellar material in which the gas is primarily in molecular form. Some nebulae are, however, associated with the final stage of stellar evolution. These are the planetary nebulae. Finally, some nebulae are detected only because they have an extremely high abundance of dust. These are the dark nebulae that absorb light from objects that lie behind them, and appear as dark 'holes' against the starry background.

nova A star that suddenly becomes 10 000 to 100 000 times as bright in a very short period of time, of between a few hours and a day, and which therefore appears to be a new star. It slowly regains its initial brightness over a period of several months or years.

nucleosynthesis The series of processes that lead to the formation of the chemical elements that make up the universe. The light elements hydrogen and helium preponderate (making up about 97% of the mass in the universe). This abundance cannot be explained through the nucleosynthesis that occurs in the centres of stars. In fact, the helium produced in stars from hydrogen through thermonuclear fusion reactions is consumed in subsequent fusion reactions that produce the heavier elements. The lightest elements, hydrogen, deuterium, and helium were probably synthesized in the initial, extremely hot and dense phase of the universe. The nucleosynthesis of heavier elements takes place within stars during the different stages of their evolution.

Olbers paradox The paradox described by H. W. Olbers in 1826, which contrasts an observational fact (the darkness of the sky at night) with the classical conception of a Euclidean universe, that is infinite, static and uniformly populated with stars, and in which calculation shows that the whole of the night sky should appear uniformly bright.

open cluster A loose grouping of a few tens to some hundreds of stars, originally born together in a single region of space, but which will eventually disperse with the passage of time.

open universe A model of an evolutionary universe that is permanently expanding.

parsec (from parallax second, abbr. pc) A unit of distance used in astronomy and corresponding to the distance at which a star has an annual parallax of 1 arcsecond.

planet A non-luminous celestial body that orbits a star, in particular, around the Sun. The current theory of the formation of planets (from a rotating nebula of gas and dust that progressively flattens into a disk, and within which larger and larger objects condense) suggests that an exceptionally large number of stars are likely to be accompanied by planets.

planetesimal Small solid bodies that result from the localized condensation of matter within the nebula surrounding a young star. Their subsequent development leads to the formation, by accretion, of planets.

plasma A gas in which all the atoms are ionized, and which therefore consists of a mixture of free electrons and atomic nuclei. Almost all the matter in the universe is in the form of plasma: extremely dense plasmas are found in the interiors of stars, whereas the solar corona, the solar wind, and nebulae of ionized hydrogen, etc. are highly tenuous plasmas.

protogalaxy A cloud of gas from which a galaxy forms through gravitational contraction.

protoplanet A planet in the process of formation through gravitational contraction, within the disk of gas and dust surrounding a star, after the latter has condensed deep inside a nebula.

pulsar (from pulsating star) A source of electromagnetic radiation (most often at radio wavelengths), charaterized by extremely short bursts, which recur at exceptionally regular intervals.

quasar (from quasi-stellar astronomical radio-source) An object of stellar appearance and exceptionally high luminosity, whose spectrum exhibits emission lines subject to a very high redshift. It is believed nowadays that quasars own their exceptional luminosity to the presence, at the centre of the galaxy of which they form the nucleus, of an extremely massive black hole (up to 100 million times the mass of the Sun). Entire stars would be sucked in, disintegrated, and would feed an immense inflow of gas.

redshift A spectral shift towards the red (i.e. towards longer wavelengths).

relativistic 1. Relating to the theory of relativity. **2.** Said of a particle that is moving extremely rapidly, with a velocity v, such that the ratio v/c (where c is the speed of light) is sufficiently large that the effects predicted by Einstein's theory of relativity are detectable.

relativity, theory of A set of theories that maintain that there are equivalent reference frames for the description of phenomena, and that the measurements relative to any one frame may be deduced following certain transformations (specific to each theory) from the same measurements made relative to another frame, with the laws of physics that express the relationships between these measurements resting invariant. General relativity thus links mechanics with gravitation. The action of gravity is described as a curvature of space created by the presence of mass.

space-time The concept of a multi-dimensional domain within which it is possible to place events and describe their mutual relationships using the co-ordinates of space and time. This concept derives from the observation of the invariance of the speed of light, regardless of the motion of the source that emits it, or of the observer. Space-time allows all observers present in the universe to describe reality independently of their relative motion. According to the theory of general relativity, gravitation is a curvature of space-time.

spectral shift The difference between the position in wavelength of the lines of an element in the spectrum of an object, and their position in a reference spectrum obtained on Earth. The shift is said to be towards the red (redshift) or towards the blue (blueshift) depending on whether the lines observed are displaced towards longer or shorter wavelengths, respectively. It may be caused by the Doppler effect or a difference in gravitational potential.

Steady State theory A cosmological theory proposed in 1948 by the astrophysicists Herman Bondi, Thomas Gold, and Fred Hoyle, in which the universe preserves the same overall appearance for all observers at all times.

supercluster A cluster of clusters of galaxies. Superclusters are the fundamental, large-scale structures in the universe.

supergiant An extremely luminous type of star, with an extremely large radius and very low density.

supernova A massive star that has reached an advanced stage in its evolution, which explodes and appears exceptionally bright for a short period.

variable star A star whose apparent magnitude varies over the course of time. The first variable star to be scientifically studied was the brilliant supernova of 1572, which was carefully observed by Tycho Brahe. The stars known as pulsating variables are stars that alternately expand and contract as a result of internal instabilities. This category includes the Cepheids, the *RR Lyrae* stars, and the *Mira* variables.

white dwarf A star with a relatively high surface temperature (approximately 10 000 K) but of very low luminosity (about one thousandth that of the Sun). Its very small radius, similar to that of the Earth, and its mass, comparable with that of the Sun, combine to produce a very high density of the order of one tonne per cm^3, and its matter is degenerate. The white-dwarf stage represents the final state in the evolution of low-mass stars.

BIBLIOGRAPHY

Arp, Halton, *Quasars, Redshifts and Controversies* Cambridge University Press, Cambridge, 1988

Audouze, Jean and Israel, Guy, *Cambridge Atlas of Astronomy:* 3rd edn., Cambridge University Press, Cambridge, 1994

Barrow, John D., *Theories of Everything,* Oxford University Press, Oxford 1990; Vintage, London, 1991

Barrow, John D. and Tipler, F., *The Anthropic Cosmological Principle,* Oxford University Press, Oxford, 1986

Combes, Françoise, Boisse, Patrick, Mazure, Alain and Blanchard, Alain, *Galaxies and Cosmology,* Springer-Verlag, Heidelberg, 1995

Contopoulos, Georgios and Kotsakis, Dimitrios, *Cosmology: The Structure and Evolution of the Universe,* Springer-Verlag, Heidelberg, 1987

Davies, Paul, *The New Physics,* Cambridge University Press, Cambridge 1989; (pbk 1992)

Davies, Paul, *About Time,* Viking, London, 1995

Einstein, Albert and Infield, Leopold, *The Evolution of Physics,* Cambridge University Press, Cambridge, 1938

Ferris, Tim, *Galaxies,* Stewart, Tabori and Chang, New York, 1989

Ferris, Tim, *The Red Limit,* Transworld, London, 1979

Flin, Piotr and Duerbeck, Hilmar, *Morphological Cosmology,* Springer-Verlag, Heidelberg, 1988

Guth, Alan, *The Inflationary Universe,* Jonathan Cape, London 1997; Vintage, London, 1998

Harrison, Edward, *Cosmology: The Science of the Universe*, 2nd edn., Cambridge University Press, Cambridge, 1999

Harrison, Edward, *Darkness at Night: A Riddle of the Universe,* Harvard University Press, Cambridge, Mass., 1989

Harrison, Edward, *Masks of the Universe,* Macmillan, 1985

Hawking, Stephen, *A Brief History of Time,* Bantam Press, London, 1988

Hoskin, Michael, *The Cambridge Illustrated History of Astronomy,* Cambridge University Press, Cambridge, 1997

Hey, Tony and Walters, Patrick, *Einstein's Mirror,* Cambridge University Press, Cambridge, 1997

Kippenhahn, Rudolf, *Light from the Depths of Time,* Springer-Verlag, Heidelberg, 1987

Krauss, Lawrence, *The Fifth Essence,* Basic Books, New York, 1989, (Hutchinson Radius, 1989); Vintage, London, 1990

Lena, Pierre, *Observational Astrophysics,* Springer-Verlag, Heidelberg, 1988

Longair, Malcolm, Our Evolving Universe, Cambridge University Press, Cambridge, 1997

Malin, David and Murdin, Paul, *Colours of the Stars,* Cambridge University Press, Cambridge, 1984

North, John, *The Fontana History of Astronomy and Cosmology,* Fontana, London, 1994

Novikov, Igor, *Black Holes and the Universe,* Cambridge University Press, Cambridge, 1995

Padmanabhan, T., *After the First Three Minutes,* Cambridge University Press, Cambridge, 1998

Pagels, Heinz R., *Perfect Symmetry,* Simon & Schuster, New York, 1985; Penguin, London, 1992

Rees, Martin, *Before the Beginning,* Simon & Schuster, London, 1997; Touchstone, London, 1998

Rowan-Robinson, Michael, *Cosmology,* 3rd edn., Oxford University Press, 1996

Rowan-Robinson, Michael, *Ripples in the Cosmos,* W.H. Freeman, New York 1993

Sandage, Allan, *The Hubble Atlas of Galaxies,* Carnegie Institution, Washington D.C., 1961

Silk, Joseph, *The Big Bang,* W.H. Freeman, New York, 1980

Thorne, Kip, *Black Holes and Time Warps: Einstein's Outrageous Legacy,* Norton, New York, 1994; Picador, London, 1994

Weinberg, Steven, *Dreams of a Final Theory,* Vintage, London, 1993

Weinberg, Steven, *The First Three Minutes,* Andre Deutsch, London, 1977 and Basic Books, New York, 1988

PHOTOGRAPHIC CREDITS

ABBREVIATIONS:

AAO = Anglo-Australian Observatory
CFHT = Canada-France-Hawaii Telescope
ESA = European Space Agency
NASA = National Aeronautics and Space Administration (Washington, D.C.)
NOAO = National Optical Astronomy Observatory
NRAO = National Radio Astronomy Observatory (Greenbank)
RAS = Royal Astronomical Society
RGO = Royal Greenwich Observatory
ROE = Royal Observatory, Edinburgh
STScI = Space Telescope Science Institute (Baltimore)
VLA = Very Large Array

PREFACE

P. 6, NASA / ESA / Ciel & Espace

A HISTORY OF COSMOLOGY

p. 8--9, S. Brunier / *Ciel & Espace*
p. 10, S. Brunier / *Ciel & Espace*
p. 11, S. Brunier / *Ciel & Espace*
p. 12, S. Brunier / *Ciel & Espace*
p. 13, S. Brunier / *Ciel & Espace*
p. 13, S. Brunier / *Ciel & Espace*
p. 13, S. Brunier / *Ciel & Espace*
p. 13, S. Brunier / *Ciel & Espace*
p. 14, S. Brunier / *Ciel & Espace*
p. 15, S. Brunier / *Ciel & Espace*
p. 16, S. Brunier / *Ciel & Espace*
p. 17, S. Brunier / *Ciel & Espace*
p. 18--19, S. Brunier / *Ciel & Espace*

THE GALAXY: AN ISLAND IN SPACE

p. 20--21, A. Fujii / *Ciel & Espace*
p. 22, A. Fujii / *Ciel & Espace*
p. 23, ROE / AAO / D. Malin / *Ciel & Espace*
p. 24, ROE / AAO / D. Malin / *Ciel & Espace*
p. 25, ROE / AAO / D. Malin / *Ciel & Espace*
p. 26, NASA / *Ciel & Espace*
p. 27, ROE / AAO / D. Malin / *Ciel & Espace*
p. 28, S. Binnewies / P. Riepe / B. Schröter / H. Tomsik / *Ciel & Espace*
p. 29, S. Binnewies / P. Riepe / B. Schröter / H. Tomsik / *Ciel & Espace*
p. 29, S. Binnewies / P. Riepe / B. Schröter / H. Tomsik / *Ciel & Espace*
p. 30--31, AAO / D. Malin / Ciel & Espace

A THOUSAND GENERATIONS OF STARS

p. 32--33, AAO / D. Malin / *Ciel & Espace*
p. 34, AAO / D. Malin / *Ciel & Espace*
p. 35, ROE / AAO / D. Malin / *Ciel & Espace*
p. 36, AAO / D. Malin / *Ciel & Espace*
p. 37, ROE / AAO / D. Malin / *Ciel & Espace*
p. 38--39, NASA / ESA / STScI / *Ciel & Espace*
p. 38, NASA / ESA / STScI / *Ciel & Espace*
p. 39, NASA / ESA / STScI / *Ciel & Espace*
p. 40, NASA / ESA / STScI / *Ciel & Espace*
p. 41, *top*, NASA / ESA / STScI / *Ciel & Espace*
p. 41, *bottom*, NASA / ESA / Ciel & Espace
p. 42, *top*, AAO / D. Malin / *Ciel & Espace*
p. 42, *bottom*, T. Gregory / CFHT / *Ciel & Espace*
p. 43, NASA / ESA / STScI / *Ciel & Espace*
p. 44, NASA / ESA / STScI / *Ciel & Espace*
p. 45, AAO / D. Malin / *Ciel & Espace*
p. 46--47, NASA / ESA / STScI / *Ciel & Espace*

THE NEXT SUPERNOVA

p. 48--49, AAO / D. Malin / *Ciel & Espace*
p. 50, NASA / ESA / *Ciel & Espace*
p. 51, A. Fujii / *Ciel & Espace*
p. 52, A. Fujii / *Ciel & Espace*
p. 53, IAC / R.G.O. / D. Malin / *Ciel & Espace*
p. 54, NASA / ESA / STScI / *Ciel & Espace*
p. 55, top, AAO / D. Malin / *Ciel & Espace*
p. 55, bottom, NASA / ESA / STScI / *Ciel & Espace*
p. 56, AAO / D. Malin / *Ciel & Espace*
p. 57, AAO / D. Malin / *Ciel & Espace*
p. 58, *top*, AAO / D. Malin / *Ciel & Espace*
p. 58, *bottom*, NOAO / *Ciel & Espace*

p. 59, NOAO / *Ciel & Espace*
p. 60, ROE / AAO / D. Malin / *Ciel & Espace*
p. 61, *top*, IAC / RGO / D. Malin / *Ciel & Espace*
p. 61, *bottom*, AAO / D. Malin / *Ciel & Espace*
p. 62--63, R. Kirchner / CFA / STScI / *Ciel & Espace*

PLANETS BY THE BILLION?

p. 64--65, NASA / ESA / STScI / *Ciel & Espace*
p. 66, NASA / ESA / STScI / *Ciel & Espace*
p. 67, NASA / ESA / STScI / *Ciel & Espace*
p. 68, NASA / ESA / STScI / *Ciel & Espace*
p. 69, NASA / ESA / STScI / *Ciel & Espace*
p. 70, A-M. Lagrange / D. Mouillet / Obs de Grenoble / ESO
p. 71, ESA / Isocam
p. 72, NASA / ESA / STScI / *Ciel & Espace*
p. 73, NASA / ESA / STScI / *Ciel & Espace*
p. 74, NASA / ESA / STScI / *Ciel & Espace*
p. 75, NASA / ESA / STScI / *Ciel & Espace*
p. 76--77, A. Fujii / *Ciel & Espace*

THE ENIGMA AT THE HEART OF THE MILKY WAY

p. 78--79, NASA / COBE / *Ciel & Espace*
p. 80, ESO / *Ciel & Espace*
p. 81, NOAO / *Ciel & Espace*
p. 82, *top*, MPIR / *Ciel & Espace*
p. 82, *bottom*, NRAO / VLA / *Ciel & Espace*
p. 83, *top*, NRAO / VLA / *Ciel & Espace*
p. 83, *bottom*, NRAO / VLA / *Ciel & Espace*
p. 84, NRAO / VLBA / *Ciel & Espace*
p. 85, CFHT / PUEO / *Ciel & Espace*
p. 86--87, NASA / *Ciel & Espace*

A SEA OF GALAXIES

p. 88--89, AAO / D. Malin / *Ciel & Espace*
p. 90, AAO / D. Malin / *Ciel & Espace*
p. 91, AAO / D. Malin / *Ciel & Espace*
p. 92, ROE / AAO / D. Malin / *Ciel & Espace*
p. 93, AAO / D. Malin / *Ciel & Espace*
p. 94, R. Schild / *Ciel & Espace*
p. 95, NASA / ESA / STScI / *Ciel & Espace*
p. 96, AAO / D. Malin / *Ciel & Espace*
p. 97, AAO / D. Malin / *Ciel & Espace*
p. 98--99, AAO / D. Malin / *Ciel & Espace*

THE ARCHITECTURE OF THE UNIVERSE

p. 100--101, ROE / AAO / D. Malin / *Ciel & Espace*
p. 102, I.A.C. / RGO / D. Malin / *Ciel & Espace*
p. 103, A. Maury / TESCA / OCA
p. 104, AAO / D. Malin / *Ciel & Espace*
p. 105, AAO / D. Malin / *Ciel & Espace*
p. 106, NOAO / *Ciel & Espace*
p. 107, Blaise Canzian / NOAO / *Ciel & Espace*
p. 108, *top*, NASA / ESA / STScI / *Ciel & Espace*
p. 108, *bottom*, NASA / ESA / STScI / *Ciel & Espace*
p. 109, *top*, F. Mirabel / CEA / *Ciel & Espace*
p. 109, *bottom*, F. Mirabel / CEA / *Ciel & Espace*
p. 110, AAO / D. Malin / Ciel & Espace
p. 111, *top*, AAO / D. Malin / *Ciel & Espace*
p. 111, *bottom*, AAO / D. Malin / *Ciel & Espace*
p. 112--113, Oxford University / *Ciel & Espace*

THE BIG BANG, AND THE HISTORY OF THE UNIVERSE

p. 114--115, NOAO / *Ciel & Espace*
p. 116, W. Baum / NASA / ESA / STScI / *Ciel & Espace*
p. 117, W. Baum / NASA / ESA / STScI / *Ciel & Espace*
p. 118, N. Tanvir, University of Cambridge / *Ciel & Espace*
p. 119, *top*, N. Tanvir, University of Cambridge / *Ciel & Espace*
p. 119, bottom, N. Tanvir, University of Cambridge / NASA / *Ciel & Espace*
p. 120, *top*, NASA / ESA / STScI / *Ciel & Espace*
p. 120, *bottom*, NASA / ESA / STScI / *Ciel & Espace*

p. 121, NASA / ESA / STScI / *Ciel & Espace*
p. 122, ESO / PSS / *Ciel & Espace*
p. 123, NOAO / *Ciel & Espace*
p. 124, NASA / ESA / STScI / *Ciel & Espace*
p. 125, ESO / *Ciel & Espace*
p. 126, ESO / *Ciel & Espace*
p. 127, O. Le Fèvre / MOS-SIS / CFHT / *Ciel & Espace*
p. 128--129, COBE / NASA / *Ciel & Espace*

GRAVITATIONAL LENSES

p. 130--131, J-P. Kneib / NASA / ESA / STScI / *Ciel & Espace*
p. 132, OMP / CFHT / *Ciel & Espace*
p. 133, CFHT / *Ciel & Espace*
p. 134, *top*, J-P. Kneib / NASA / ESA / STScI / *Ciel & Espace*
p. 134, *bottom*, NRAO / VLA / Ciel & Espace
p. 135, *top*, CFHT / Ciel & Espace
p. 135, *centre*, CFHT / Ciel & Espace
p. 135, *bottom*, H. Arp / NASA / *Ciel & Espace*
p. 136, *top*, RAS / *Ciel & Espace*
p. 136, *bottom*, NRAO / VLA / *Ciel & Espace*
p. 137, *top*, Keck / CARA / *Ciel & Espace*
p. 137, *bottom*, Keck / CARA / *Ciel & Espace*
p. 138, *top*, J-P. Kneib / NASA / ESA / STScI / *Ciel & Espace*
p. 138, *bottom*, J-P. Kneib / NASA / ESA / STScI / *Ciel & Espace*
p. 139, J-P. Kneib / NASA / ESA / STScI / *Ciel & Espace*
p. 140--141, J-P. Kneib / NASA / ESA / STScI / *Ciel & Espace*

LE MYSTERY OF THE MISSING MASS

p. 142--143, O. Le Fèvre / G. Luppino / CFHT / *Ciel & Espace*
p. 144, O. Le Fèvre / F. Hammer / CFHT / *Ciel & Espace*
p. 145, B. Fort / NASA / ESA / STScI / *Ciel & Espace*
p. 146, NOAO / Sackett & Al / *Ciel & Espace*
p. 147, B. Fort / Y. Mellier / CFHT / *Ciel & Espace*
p. 148, *top*, ROSAT / *Ciel & Espace*
p. 148, *bottom*, S.D.M. White / U.G. Briel / J.P. Henry / ROSAT
p. 149, *top*, Observatoire de Marseille / VLA / N.R.A.O
p. 149, *bottom*, P. Ho / CFA / K.Y. Lo / University of Illinois / M.S. Yun / Caltech
p. 150--151, NASA / ESA / STScI / *Ciel & Espace*

SEARCHING FOR THE ULTIMATE

p. 152--153, R. Williams / HDF / STScI / NASA / *Ciel & Espace*
p. 154, R. Williams / HDF / STScI / NASA / *Ciel & Espace*
p. 155, R. Williams / HDF / STScI / NASA / *Ciel & Espace*
p. 156, NASA / ESA / STScI / *Ciel & Espace*
p. 157, R. Williams / HDF / STScI / NASA / *Ciel & Espace*
p. 158--159, Luppino & Kaiser / U.H. / *Ciel & Espace*
p. 160--161: NASA / ESA / STScI / *Ciel & Espace*
p. 162, *top*, G. Lelièvre / CFHT / *Ciel & Espace*
p. 162, *bottom*, NASA / ESA / STScI / *Ciel & Espace*
p. 163, *top*, NASA / ESA / STScI / *Ciel & Espace*
p. 163, *bottom*, Keck / CARA / *Ciel & Espace*
p. 164, J-P. Kneib / NASA / ESA / STScI / *Ciel & Espace*
p. 165, Keck / CARA / *Ciel & Espace*
p. 166--167, R. Williams / HDF / STScI / NASA / *Ciel & Espace*

TOWARDS THE COSMOLOGICAL HORIZON

p. 168--169, CERN
p. 170, CERN
p. 171, CERN
p. 172--173, CERN
p. 172, *bottom*, NRAO / VLA / *Ciel & Espace*
p. 173, *bottom*, NRAO / *Ciel & Espace*

p. 174, ESO / *Ciel & Espace*
p. 175, Keck / CARA / *Ciel & Espace*
p. 176, Keck / CARA / *Ciel & Espace*
p. 177, RGO / RAS / *Ciel & Espace*
p. 178, E. Turner / J-P. Kneib / NASA / ESA / STScI / *Ciel & Espace*
p. 179, E. Turner / J-P. Kneib / NASA / ESA / STScI / *Ciel & Espace*
p. 180--181: M. Franx / NASA / ESA / STScI

APPENDICES

p. 182--183, S. Brunier / *Ciel & Espace*
p. 184, S. Brunier / *Ciel & Espace*
p. 185, S. Brunier / *Ciel & Espace*
p. 186, S. Brunier / *Ciel & Espace*
p. 187, S. Brunier / *Ciel & Espace*
p. 188, S. Brunier / *Ciel & Espace*
p. 189, S. Brunier / *Ciel & Espace*
p. 190, NASA / *Ciel & Espace*
p. 191, NASA / *Ciel & Espace*
p. 192, S. Brunier / *Ciel & Espace*
p. 193, S. Brunier / *Ciel & Espace*
p. 194, S. Brunier / *Ciel & Espace*
p. 195, S. Brunier / *Ciel & Espace*
p. 196, S. Brunier / *Ciel & Espace*
p. 197, S. Brunier / *Ciel & Espace*
p. 199, NASA / ESA / STScI / *Ciel & Espace*
p. 200--201, Olivier Hodasava / *Ciel & Espace*
p. 202--203, Manchu / *Ciel et Espace*
p. 204--205, Olivier Le Fèvre / Olivier Hodasava / *Ciel & Espace*
p. 206, Olivier Hodasava / *Ciel & Espace*
p. 207, Olivier Hodasava / *Ciel & Espace*

INDEX